best of british
STEAM ENGINES

Ian Mackenzie

best of british

STEAM ENGINES

First published in the UK in 2015

www.g2books.co.uk

Printed and bound in Europe

ISBN 978-1-78281-886-1

building up to 1913 meant by then the total number was 289, numerically the largest locomotive class of the GER.

The sturdy reliability of the J15s meant that as late as 1959, 33 were still in service with British Railways. Withdrawal of the final four coincided with the elimination of steam in East Anglia, in September 1962. One of those last four, BR No 65462, has survived into preservation.

Though largely unsung during their working lives the near 80 year long career of these little engines was a tribute to Thomas Worsdell's original design and the workmanship of Stratford Locomotive Works.

ABOVE Great Eastern Railway Y14 / LNER Class J15 0-6-0 No. 65462 with demonstration freight train over River Avill Bridge

Caledonian Railway

CALEY 4-2-2 No. 123

RIGHT
Caledonian railway no 123 at Glasgows transport museum

The famous single, Caley 123, was a lone engine specially built by Neilson & Company for the 1886 Edinburgh International Exhibition where it was awarded a gold medal.

The invention of the steam sander in 1885 provided a fillip for the final generation of single wheeler passenger express engines in the last years of the nineteenth century.

Although the detailed design work was undertaken by Neilson the engine shows the clear influence of the Caledonian Railway's famous locomotive chief, Dugald Drummond, and after the Exhibition the engine was acquired by the Caledonian for £2,600.

In its blue livery, 123 was a splendid advertisement for both Nielson's and the Caledonian, particularly after its spectacular performance in the 1888 "Railway Races to the North". Between the 6th and 13th August, 123 was responsible for the final 101 miles from Carlisle to Edinburgh. Each day the engine arrived comfortably within the 112 minute timing, saving nine-and-a half minutes on the final day.

Something of a pet, 123 survived grouping in 1923, hauling inspection saloons and piloting the royal train until, in 1930, it was returned to public service on the route between Perth and Dundee.

The engine toured as a static exhibit until 1959 when it was restored to working order and then ran enthusiast railtours in Scotland during the 1960s.

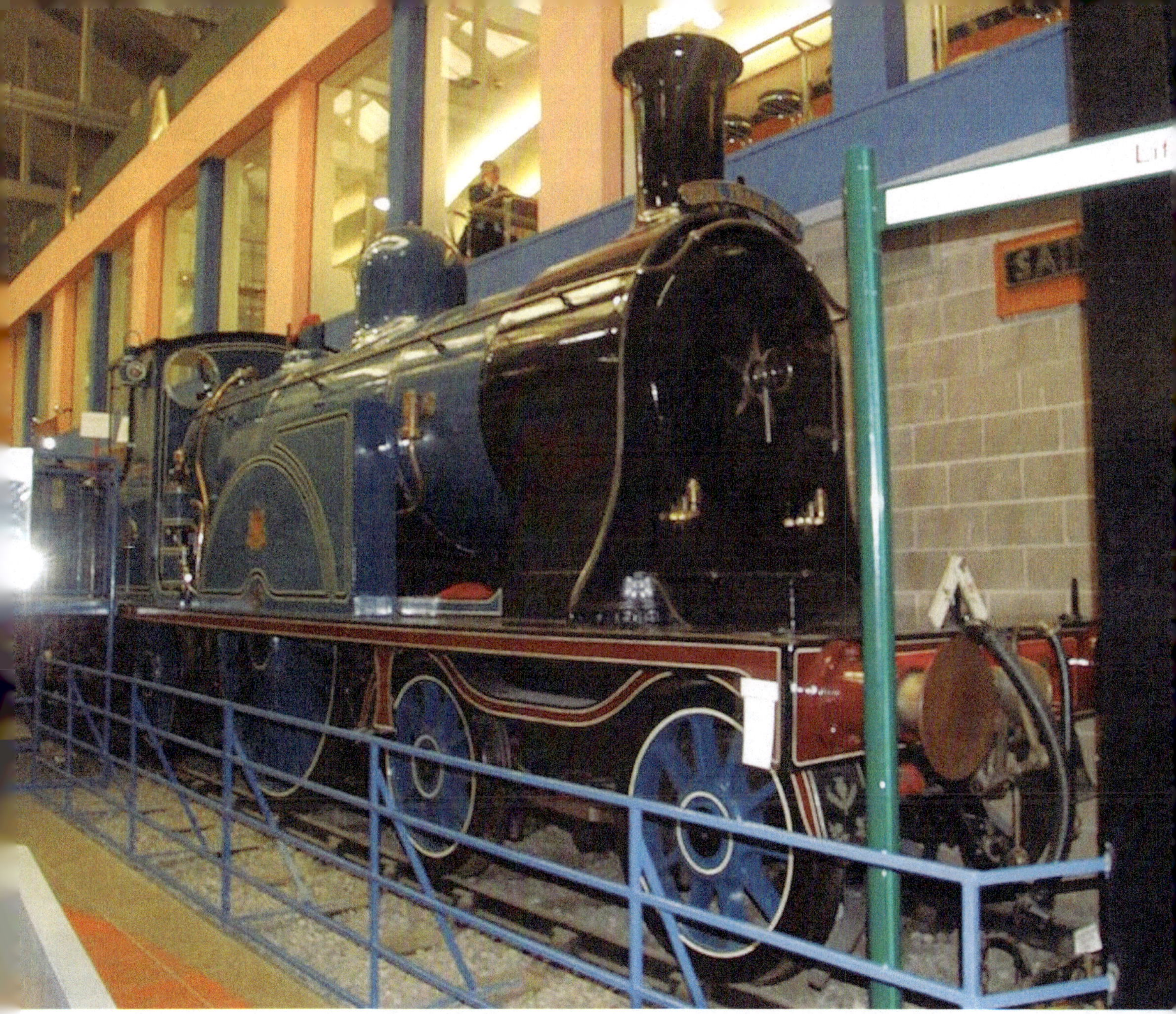

South of the border, one of its last tours, in September 1963, was piloting another famous Drummond engine, T9 number 120 of the LSWR, on the Bluebelle Special from Victoria to Horsted Keynes on the Bluebell Railway.

6

London & South Western Railway

02 0-4-4 TANK [1889]

BELOW
Isle of Wight Steam Railway, a train hauled by Class O2 locomotive no 24 "Calbourne

First introduced in 1889, these reliable engines outlived their normal lifespan when they became the major motive power on the Isle of Wight from 1923 to 1966.

Having previously being wedded to the 4-4-0 wheel arrangement for his suburban locomotives, William Adams introduced his first 0-4-4, the T1 tank, for the LSWR in 1888. With its 5' 7" wheels, this engine was specifically designed for fast acceleration and speed on the London suburban services.

The following year, 1889, saw the introduction of the O2, an 0-4-4T smaller and lighter than the T1. This engine, with its 4' 10" wheels was designed to fulfil a wide range of duties and the 60 engines built between 1889 and 1995 were allocated all over the South

Western: from the Surrey suburbs to the western outposts of Devon and Cornwall.

Following the introduction of Drummond's larger M7 0-4-4T, O2s were cascaded to humbler duties and would undoubtedly have succumbed to obscurity and scrapping but for the testing of two of the class, 206 and 211, on the Isle of Wight in 1923.

The O2s proved admirable and by 1936 the Southern Railway had shipped 21 engines to the Isle of Wight. For service there, the engines were fitted with Westinghouse air brakes and had their bunkers extended to take three tons of coal as shown in the picture of No 24 in British Railways livery near the end of its working life.

ABOVE Calling at Esplanade station on a the 14.35 Ryde Pier Head to Cowes is No. W24 'Calbourne'

7

LNWR ("Jumbos")

HARDWICKE No. 790 [1892]

Described by railway journalist Charles Rous-Marten as "wonderful little engines", Francis Webb's Improved Precedents justly deserved their nickname "Jumbos": little engines with the strength of an elephant.

During the late nineteenth century the London and North Western Railway was for a time the world's largest joint-stock company and concentrated on building shareholder value in a way entirely familiar with business practice today. By running lightweight trains with simple engines and keeping average journey speeds at 40mph the shareholders reaped a competitive dividend.

Francis Webb's Precedent Class, was first introduced in 1874, and admirably met the business criteria: cheap to build, and smaller and lighter than their contemporaries on other railways.

The success of the Precedents persuaded Webb to standardise the locomotive fleet and, from 1887, start rebuilding earlier locomotive classes of engines as "Improved Precedents". Although nominally a rebuild of a Newton Class locomotive dating back to 1873, Hardwicke was, to all intent and purpose, a new engine but, based on a design going back to 1874, starting to look small and old-fashioned when compared to the express engines of other railway companies.

During the summer of 1895 the east coast railway companies accelerated their overnight services to Scotland and revived the "Railway Races to the North" of 1888. The LNWR in conjunc-

tion the Caledonian Railway responded by re-timing its sleeping car service to Aberdeen. Over the weeks the racing intensified, reaching a climax on the night of 22nd/23rd August. Hardwicke was in charge of the leg from Crewe to Carlilse and that night covered the 141 ½ miles on 126 minutes, an average speed of 67.3mph including topping Shap Summit at 65.3mph. Unconfirmed reports suggest the engine's top speed on the jourmey was 96mph. That performance secured Hardwicke's name for posterity and ultimately secured its preservation by the LMS on withdrawal in 1932.

ABOVE L&NWR "Jumbo" class 2-4-0 No. 790 Hardwicke in steam and working agin under her own power

8

Highland Railway

JONES GOODS 4-6-0 [1894]

The 4-6-0 wheel made its first appearance in the USA in 1847 and had been produced by British manufacturers for overseas use since the 1860's. The Jones Goods became famous as it was the first 4-6-0 to run on Britain's rails.

The Highland Railway's 144 mile mainline from Perth to Inverness was a series of modest climbs and two particularly taxing summits: Druimauchdar, which at 1,484 feet is the highest point on the British railway network and Dava Moor, 1,052 feet. Enginemen referred simply and succinctly to the route as "the Hill".

David Jones, a Manchester born engineer, succeeded William Stroudley as Locomotive Superintendent on the Highland Railway in 1870. He designed a series of 4-4-0 engines, including the Dukes and Straths, which proved more than adequate for traffic on "the Hill" during the 1870s and 1880s. In 1890 the Forth Bridge opened with a substantial increase in traffic, heavier trains and the need for some more powerful engines on the Highland Railway.

Instead of opting for larger 4-4-0s, David Jones instead took the radical step of designing and commissioning a 4-6-0 and the 15 engines built by Sharp Stewart for the Highland railway were the first 4-6-0s to run in Britain. At the time their introduction caused considerable stir, mainly because the Highland Railway was a minor railway with a relatively unknown locomotive superintendent.

The Highland was a small railway and the class totalled just fifteen engines,

built and delivered by Sharp, Stewart and Company between September and November 1894.

Withdrawal of the locomotives took place between 1929 and 1940 but No 103 (LMS No 17916) was set aside for preservation by the LMS in 1934. It was restored to working order by British Railways in 1959 and then spent several years operating enthusiasts' specials as well as appearing in the 1965 film Those Magnificent Men in their Flying Machines.

ABOVE Highland Railway Jones Goods Class No 103 at the Riverside Museum

The Caledonian Railway

4-4-0 DUNALASTAIRS [1897]

Belgian State Railways were so impressed by the Dunalastair exhibited in Brussels in 1897 that over the next 10 years they built 720 for use in Belgium, far outnumbering those built for the Caledonian.

From humble beginnings John McIntosh started work on the railways in 1860 and progressed through the ranks: from fireman, engine driver, running foreman, Chief Running Superintendent and finally Locomotive Superintendent of the Caledonian Railway following the sudden death of the previous incumbent, John Lambie, in 1895.

McIntosh's first design as Locomotive Superintendent was a 4-4-0 passenger express, No.721, Dunalastair, a big, big boilered engine which demonstrated that many years as a practical engineman had taught McIntosh what an engine needed "on the road".

Following the success of 721, the class, now known as Dunalastairs after the first engine, was quickly expanded to 15 engines which were immediately entrusted to the Caledonian's section, Glasgow to Carlisle, of the prestigious Anglo-Scottish "Corridor" express.

A more demanding task was the "Tourist", 183 ½ miles from Carlisle to Forfar via Stirling and Perth. Without any water troughs on the Caledonian, the engines were soon fitted with eight-wheeled tenders carrying 4,125 gallons of water, as illustrated here.

Over the next few years McIntosh intro-

duced variants of the Dunalastairs each marking a progressive increase in size over its predecesser: 15 Dunalastair IIs in 1898, three Dunalastair IIIs in 1899 and in 1904, 19 Dunalastair IVs.

The final development, in 1910, was to build a Dunalastair IV, the 20th, with a Schmidt-type superheater. Comparative tests against a saturated Dunalastair IV registered dramatic coal and water savings and a further four superheated Dunalastair IVs were ordered in 1911.

Although no British Dunalastair has survived, in a curious historical footnote one of the larger variants built by Belgium's state railway has been preserved. SNCB Class 18 No 18051 can be seen at the SNCB Railway Museum in Treignes faithfully restored in the Caledonian blue livery chosen for these McIntosh designed engines.

ABOVE Preserved SNCB Class 18 Dunalastair type No 18051

10

London and South Western Railway

M7 0-4-4 TANK [1897]

Before he came to the LSWR Dugald Drummond had established a reputation as a gifted locomotive engineer, a reputation enhanced by his first design for the LSWR – the M7 tank engines.

When Dugald Drummond joined the LWSR in 1895 he was already an experienced engineer with a reputation for designing effective and elegant engines. He had been Locomotive Superintendent on the Highland Railway, the North British Railway and the Caledonian Railway. The grace and proportion of his engine designs probably owed much to the mentoring he had received while working for both Samuel Johnson and William Stroudley.

His first designs for the LSWR, the M7 0-4-4T and the Class 700 0-6-0 freight engine maintained that reputation.

Larger than the LSWR's O2 0-4-4T and in size harking back to Adam's T1 0-4-4T of 1888, the M7s were primarily introduced for suburban service. Showing a turn for speed, however, the running department started using them on secondary express passenger duties and soon they could be seen on services to Portsmouth and Bournemouth and semi-fasts between Exeter and Plymouth.

Underlining the success of the design, M7s continued to be built up to 1911 by which time the class numbered 105 engines. At the time of grouping in 1923 this was the Southern Railways largest single class of tank locomotives.

Following electrification the M7s contin-

ued their role as reliable branch-line locomotives with all 105 engines passing into British Railway's ownership. Withdrawals didn't start until 1957 and during the 1950s and 60s M7s could be seen all over the Southern Region: on branch-line duties, empty stock workings and operation between Waterloo and Clapham yard.

ABOVE LSWR Class M7 tank loco No. 245

South Eastern & Chatham Railway

D CLASS 4-4-0 [1901]

RIGHT South Eastern and Chatham Railway (SECR) D Class 4-4-0 locomotive no 737, built at Ashford Works in 1901

With their burnished copper chimney caps and polished dome and safety valve covers, the D Class quickly acquired the nickname "Coppertops".

By the mid 1890s the inside cylinder 4-4-0 was established as the standard locomotive for express passenger work. With modest train weights these small engines had shown themselves capable of both high speed and tackling the steepest gradients.

Apart from the coupling rods, all the motion was accommodated between the frames so that a straight footplate combined with balanced design produced clean lines traditionally associated with late Victorian British engines.

No engine epitomised this elegant, stylish look better than Harry Wainwright's D Class locomotive introduced on the South Eastern & Chatham Railway in 1901.

Design of the engine was collaboration between Locomotive, Carriage and Wagon Superintendent Harry Wainwright and his Chief Draughtsman, Robert Surtees. Wainwright dealt with the artistic aspects while Surtees took care of the mechanical details.

Although there was nothing especially innovative about the D Class, its handsome looks were reflected in its performance. Sturdy and surefooted, the engines rode well, steamed freely and were economical.

By 1907 the class totalled 51 engines, 22 built at Ashford and the remainder from outside contractors. Wainwright's

successor, Richard Maunsell, rebuilt 21 of the class with Belpaire boilers and superheaters to produce the more powerful D1s.

Most of the remaining engines passed into British Railway's ownership, the last withdrawal taking place in 1956. By then BR No. 31737 had already been nominated for preservation by the BTC and was returned to its full SECR livery at Ashford in 1959.

12

Midland Railway 4-4-0 COMPOUND No. 1000 [1901]

RIGHT Restored in all its original crimson splendour, Midland Railway 4-4-0 Compound stands outside Derby Works. Withdrawn in 1951 as BR No 41000 the engine, the first of its class, was earmarked for preservation by the British Transport Commission

In the years before the First World War the widespread use of the red liveried compounds across the Midland Railway earned them the nickname "Crimson Ramblers"

Considering Samuel Johnson's well recorded antipathy to compounding it was something of a surprise when the first two Midland Railway 4-4-0 compounds, numbers 2631 and 2632, were introduced in 1901.

The engines used the compounding system devised by Scotsman Walter Smith, chief draughtsman at the North Eastern Railway's Gateshead works. Smith's system used a single high-pressure cylinder between the frames with two low-pressure outside cylinders. Although Johnson knew Smith well, they had worked together at the Great Eastern's Stratford works, it is more likely that the main champion for the development and building of the compounds was Derby works Manager Richard Deeley. Deeley took over as CME when Johnson retired in 1904, by which time five Johnson compounds had been built.

Deeley introduced his improved version of the compound in 1905: an engine with increased boiler pressure, larger firebox and cab and simplified compound controls. By the time Deeley resigned in 1909, 45 compounds were in service, all of which were superheated from 1913 onwards.

As a testament to the success of the compounds the LMS built a further 195 between 1924 and 1932 making the class of

240 engines the largest series of three cylinder compounds anywhere in the world.

All the Midland engines had been withdrawn by 1953 while the LMS built engines remained in service until 1961. The first engine of the class, now numbered 41000, was identified as an engine to be preserved for the nation when withdrawn in 1953.

Although now a static exhibit at York, no 1000 was restored to working order on two occasions following preservation, first from 1959 – 1963 and again from 1976 – 1983.

13

Great Northern Railway

CLASS C1 ATLANTIC No. 251 [1902]

By the end of the nineteenth century it was clear that development of the classic inside cylindered 4-4-0 had reached a plateau and rapidly increasing train-weights demanded something more powerful. Henry Ivatt's Atlantic classes on the GNR represented a significant and innovative design step change.

The need for locomotive design change was very clear to Henry Ivatt, appointed Locomotive Superintendent of the Great Northern Railway in 1896, where he could see that current train loadings were taxing both the famous "singles" and the newer 4-4-0s.

Ivatt saw the way to greater power in simple terms: larger boilers with greater steam raising power but integrated with proportionate cylinders and steam circuit. It was natural for Ivatt to turn to the 4-4-2 Atlantic design established in America by the Baldwin Locomotive Company in 1895. With an extended frame and additional axle the Atlantic could support a bigger boiler and firebox and so generate greater power.

Ivatt built his first Atlantic, No. 990, named after the GNR's General Manager Henry Oakley, in 1898 and then subjected the engine to two years of intensive testing before building the first 10 of the class in 1900.

This class of Atlantics, the C2s, known later as "Small Atlantics" were merely a prelude to the introduction of Ivatt's masterpiece, the "large Atlantics", Class CI.

The prototype engine, No.251, illustrated

here, was out-shopped from Doncaster Works in December 1902. With its massive Wootten-type firebox, almost the full width of the engine, and enormous boiler, No. 251 dwarfed the C2 Atlantics. Another two years of testing the prototype confirmed its success and the Class went into series production from 1904 onwards.

All the C2 Atlantics had been withdrawn from service by 1950 but not before the pioneer, No 251, had been selected for preservation as part of the national collection.

ABOVE GNR large boilered Atlantic No 251 at Bressingham Steam Museum

14

Great Western Railway

4-4-0 CITY OF TRURO [1903]

BELOW
GWR City of Truro at Didcot

City Class engine City of Truro achieved lasted fame and subsequent preservation when it became the first engine to exceed 100mph on 9th May 1904.

Although fine mid-sized engines, there was little exceptional about Churcward's City Class engines. They represented the final development of William Dean's doubled-framed four coupled engines improved by the addition of the Swindon Standard No 4 tapered boiler.

The double frames allowed a significant increase in bearing surfaces and the possibility of high speed running without axle-boxes overheating. This and their reputation for mechanical excellence meant that a City Class engine was chosen in a deliberately staged demonstration of speed. At 9:23am on the 9th May 1904 a mail-train of five vans headed by City of Truro set off for London with mail and gold bullion unloaded from the liner Kronprinz Wilhelm, en route from New York to Hamburg.

Though not normally a passenger service, the GWR on this occasion invited Charles Rous-Martin, the famous railway commentator and speed recorder, to join the train. Some high speed running was planned and expectations were not disappointed when Rous-Marten's stopwatch recorded a time of 8 and 4/5 seconds for one quarter of a mile during the descent of Wellington bank: a speed of 102.3 mph.

Though the accuracy of this figure was subsequently surrounded in controversy it was not enough to stop City of Truro achieving immortality. But for that single event, City of Truro would have been scrapped along with its class mates in the 1930s after a relatively short 30 year working life.

Withdrawn in 1931 the engine was transferred to the LNER's York Museum but was subsequently restored to mainline running on three occasions: first in 1957, then 1985 and finally 2004. Restoration in 1904 meant a special centenary run heading the "Ocean Mail 100" on the 100th anniversary of the speed record.

ABOVE GWR 3440 City of Truro

15

The Great Western Railway LARGE PRAIRIES 2-6-2 TANK [1905]

From his original prototype of 1903 G.J. Churchward created an engine so suited to its purpose that the class continued in production with minor detail changes right up to 1949.

BELOW
GWR 6100
Class 6106

The Large Prairie tank was one of the designs G.J. Churchward selected to launch his re-equipping of the Great Western's locomotive fleet. Development started with a prototype, number 99, which was out-shopped from Swindon in 1903. After two years of tests with number 99, production of the first series of Large Prairies started in 1905.

Churchward remained as CME of the Great Western until 1922 and during his tenure Large Prairies continued to be built with progressive design improvements: larger diameter chimneys, top feed, superheating, increase in boiler pressure and an increase in bunker size. Prairies built during this period fell into two classes, the 3100 and 3150, later designated 5100 and 5100 by Churchward's successor, Charles Collet.

The 5100s could be seen all over the Great Western except around London, where Churchward's 4-4-2 County tanks were dominant. By the 1930s it was clear that something more powerful was needed to meet the demands of the newly accelerated Paddington suburban services.

Using the 5100 Class as a start point, Collett increased boiler pressure from 200 psi to 225 psi to create 60 engines of the 6100 Class like 6106 pictured here. With rapid acceleration and a high turn of speed the 6100 Class were admirably suited to the Paddington suburban work.

Although none of the larger 6100 class has survived, three 5100s have been preserved: No 5164 on the Severn Valley Railway, No 5193 on the West Somerset Railway and No 5199 on the Llangollen Railway. For operational reasons the West Somerset Railway has rebuilt 5193 as a 2-6-0 tender engine resembling a 4300 Mogul and renumbered 9351.

TOP GWR 3120 Prairie locomotive 1907

ABOVE West Somerset Railway GWR 2-6-2T 4160

16

Great Central Railway
2-8-0 FREIGHT ENGINE (LNER CLASS O4) [1911]

Service in both World Wars and overseas sales meant that these GCR 2-8-0s saw service in nine countries, including Australia where one of the last survivors was still at work with its original boiler in 1973.

BELOW
63601 drifts into Rothley station

For the 1912 opening of its new Immingham dock complex on Humberside the Great Central Railway realised it would need some new heavy freight engines. After considering 2-8-0s, 2-8-2s and 2-10-0s, John Robinson, the GCR's CME, eventually settled on a 2-8-0. The first engine of the new class was shopped out of Gorton Works in 1911.

Further building at Gorton and by outside contractors Kitson & Co and the North British Locomotive Company meant that by 1914 the GCR had 126 of these robust, reliable and uncomplicated engines.

The reliability and simplicity of the engine recommended it to the Ministry of Munitions in the First World War when there was a need to build new engines rather than borrow from Britain's railway

companies. Engines were ordered for the military, specifically the Railway Operating Division (ROD) of the Royal Engineers and to keep unemployment figures down after the war the government continued with construction of the 8Ks until by 1920 521 redundant ROD 2-8-0s were put up for sale.

Initially slow to move a variety of incentives and knock down prices meant that by 1927 all the engines had been sold, mainly to the LNER though a number ended up abroad, including China and a mining company in New South Wales. Modifications and alterations to the 400 plus engines acquired by the LNER meant that, though nominally designated class O4, by 1944 and in typical LNER style the class had been divided into 8 separate sub-classes.

The sole survivor, one of the original Robinson GCR engines, number 102 was preserved as part of the national and has now been restored to working order on, appropriately the GCR heritage railway.

ABOVE J & A Brown ROD 23 at Hexham June 1973

17

The Great Western Railway
MOGUL 2-6-0 4300 CLASS [1911]

A tentative suggestion by Harold Holcroft, one of G.J.Churchward's assitants at Swindon, to build a mixed traffic 2-6-0 resulted in the 4300 Class, so successful that by 1932 the GWR had built 342, its second largest class of locomotives.

During the summer of 1909 Harold Holcraft, a bright Swindon Works junior, had been one of a party of engineers touring railroads of the USA and Canada.

Holcroft was fascinated by the universal use there of outside cylinder 2-6-0s for both passenger and freight traffic. With poorly laid track and low average speeds of around 45 mph the type had become popular in North America because it offered greater power and adhesion than the classic American outside cylindered 4-4-0. The low average speeds were dictated by the small wheels, under 5 ft diameter, of the typical American mogul.

Holcroft reasoned that an 8 – 10 inch increase in wheel size could increase speed up to 60 mph, ideal for a mixed traffic engine in Britain. Holcroft's ideas received approval from Churchward (GWR CME, 1902 – 1922) who reportedly said:

"Very well then: get me out a 2-6-0 with 5 feet 8 inches wheels, outside cylinders, the number four boiler and bring in all the standard details you can."

And from that Britain's first truly mixed traffic locomotive was conceived.

Within a few days Holcroft had com-

pleted the outline draft. Using standard parts and relying on the design of the Large Prairie 2-6-2 of 1906 meant that most of the drawings existed and only layout diagrams had to be newly drawn. For that reason the locomotive did not need to go through the usual prototype testing stage and Churchward placed an order for 20, the first of which were delivered in 1911.

The engines proved very successful and construction of the class continued through to 1932.

Just two engines have survived into preservation. No 5322 built in 1917 and now restored to working order at the Didcot Railway Centre in its First World War guise as ROD5322 and No 9303 awaiting overhaul on the Severn Valley Railway.

ABOVE
ROD 5322 at Didcot railway centre

18

The Great Eastern Railway
4-6-0 1500 CLASS (LNER B12) [1911]

To meet the demand for additional power on the GER, newly appointed Locomotive Superintendent Stephen Holden delivered the 1500 Class 4-6-0, the first of only two new engine classes he built for the GER before retirement in 1912.

BELOW
Weybourne Station, 7th April 2007

When Stephen Holden succeeded his father James Holden as Locomotive Superintendent of the GER in 1908 the principal Norwich and Harwich expresses were handled by James Holden's 4-4-0 designs, notably his Claud Hamiltons. Increased train loadings were putting a strain on the 4-4-0s so one of Stephen Holden's first tasks as the new Locomotive Superintendent was to devise something more powerful.

The weight and length of new locomotives was restricted by the GER's historically poorly engineered permanent way and the size of its turntables. But guided by his Chief Draughtsman, Frederick Russell, Stephen Holden produced the

1500 Class, a handsome 4-6-0.

The first engine was outshopped from Stratford in 1911 and though clearly a twentieth century superheated engine, bore all the hallmarks of a nineteenth century engine: ornate styling, copper-capped chimney, extensive brass beading and elaborate fretted wheel splashers.

The engines were good performers well liked by their crews and by grouping in 1923 the GER had 70 1500 Class 4-6-0s. Between then and 1933 Nigel Gresley added further engines to the class, with boiler, firebox and valve gear changes to each new batch so that there were eventually four B12 sub-classes on the LNER. With its weight and loading gauge advantages the expanded class was available for use on other regions of the LNER and B12s could be found on the former Midland & Great Northern Joint Railway and the old GNoSR line in Scotland.

One engine has survived, LNER variant B12/3 No 8572, which is operational and can be seen on the North Norfolk Railway.

ABOVE No. 61504, one of the original ex-GE B12/1 4-6-0s designed by Holden for the GER

19

Great Central Railway

4-4-0 CLASSES

11E (LNER D10) & 11F (LNER D11) [1913]

Named after directors of the Great Central Railway the celebrated 4-4-0s of 1913 quickly became known to railway enthusiasts as "Directors".

John Robinson was appointed as Locomotive Superintendent of the GCR in July 1900, just over a year after the opening of the railway's "London Extension" to Marylebone.

He inherited a parlous locomotive situation from his predecessor, Harry Pollitt with a shortage of engines and few engines powerful enough on the London Extension to match the timings of the Midland Railway and LNWR.

The engine shortage was solved by borrowing 50 engines from other railway companies and within a year Robinson had delivered a new 4-4-0 (Class 11B), followed in 1903 by the first of his celebrated Atlantics.

Disappointed with the performance of his subsequent 4-6-0s and spurred on no doubt by the success of the LNWR 4-4-0 George the Fifth Class, Robinson reverted to the 4-4-0 format and produced a winner.

The features that made the new engines such a success were its superb boiler design, ample heating surfaces, large firebox, piston valves and, most importantly a superheater.

The engines of the new class were named after directors of the GCR, the first, N0.429, Sir Alexander Henderson, delivered from Gorton Works in 1913. The engines were an immediate success and

could match or even better the 4-6-0s on most train loadings with significantly less fuel consumption.

10 engines were built before the First World War and the class dominated the Manchester to Marylebone expresses with especially impressive performances over the 103 mile stretch from Leicester to London.

Construction resumed after the War by which time the engine had evolved to Class 11F, the "Improved Directors" with more substantial cabs and improved steam raising ability.

ABOVE Butler Henderson at Barrow Hill

London & North Western Railway
4-6-0 "CLAUGHTON" [1913]

BELOW
A Claughton running in the 1920s

Although initially demonstrating superlative test performance, daily operation of the "Claughtons" revealed all sorts of problems. Were the problems down to ruthless handling, poor construction or was the design fundamentally flawed?

By 1909, when Charles Bowen Cooke was appointed CME, the LNWR had severe motive power problems. As well as an overall shortage of engines the company had nothing comparable in size and power to the 4-4-2s and 4-6-0s being run by other railway companies.

Bowen Cooke's approach to the problem was prudent and considered. He arranged comparative trials with the latest locomotives of rival companies, including a four-cylindered Star Class 4-6-0 from the Great Western.

From this detailed research Bowen Cooke produced a superheated version of the 4-4-0 Precurser Class, followed by the 2 cylinder 4-6-0 Prince of Wales Class and, finally, his ultimate goal, a four-cylindered 4-6-0.

The first engine, No 2222, was named after the LNWR's Chairman, Sir Gilbert Claughton, and gave the class its name. On paper and in spite of the careful planning that had gone into the design the locomotive's projected performance did not look encouraging, particularly in view of the small boiler and very modest working pressure of only 175 psi.

ABOVE LMS Enlarged Claughton class, 5986 (CJ Allen, Steel Highway) 1928

Initial tests confounded these fears and dynamometer readings made it clear that the LNWR finally had an engine capable of outstanding performance. Sadly the early promise was not sustained and daily use over the next 10 years produced a litany of faults: axle boxes were prone to overheat, lubrication of the valves was poor, the coupled wheel springs suffered frequent breakages, smokebox doors warped and the shallow frames had problems with cracking. The engines became increasingly expensive to maintain and a growing embarrassment for the company.

After grouping and the appointment of William Stanier as CME of the LMS a kind of resolution of the Claughton problem was achieved. Substantial numbers were rebuilt as the new Patriot class, popularly known as "Baby Scots". But in reality, "rebuild" was an accounting fiction and, apart from some material recycling, the "new" Patriots were just that – new engines.

Neither a Claughton nor a Patriot has survived but with the enthusiastic modern trend to recreate lost historic engines, the Llangollen heritage railway has set about building a new Patriot, to be named The Unknown Warrior.

21

North Eastern Railway
T2 (LNER Q6) 0-8-0
FREIGHT LOCOMOTIVE [1913]

Vincent Raven's development of Wilson Wordsell's 0-8-0 freight engines resulted in the introduction in 1913 of the supremely successful and reliable T2s, most of which survived until the end of steam with only minor modifications.

BELOW
63995 at Sunderland shed

At the start of the twentieth century freight accounted for over 60% of the North Eastern Railway's traffic, hauled by a huge fleet of simple, sturdy 0-6-0 goods engines.

The NER had been slower than its neighbours, the LNWR and L&Y, to upgrade from 0-6-0 to more powerful 0-8-0 goods engines, but that was to change in 1901 when Gateshead Works outshopped the first of Wilson Wordsell's T Class eight coupled locomotives.

Ten T Class engine were built in 1901 and they quickly proved their worth handling 1,200 ton coal trains from Pelaw Colliery to Tyne Dock. Later variants of the class were equipped with slide valves and the

two classes were identified as T and T1 (LNER Q5/1 and Q52).

Building on the success of the Ts, Wordsell's successor, Vincent Raven, developed a bigger, more powerful version, the T2 (LNER Q6) pictured here. With a bigger boiler, superheating and piston valves the new engines were an unqualified success. Building the T2s continued up to 1921, by which time the class totalled 120 engines.

With only a few minor modifications over the ensuing years, all 120 engines passed into the ownership of British Railways. Withdrawals took place from 1963 onwards with the last 17 surviving until 1967.

Number 63395, one of the last 1967 survivors, was successfully purchased by the North Eastern Locomotive Preservation Group and restored to working order on the North York Moors Railway were it ran until withdrawal for overhaul in 1982. Getting adequate resources meant a delayed overhaul but the engine returned to service on the North York Moors Railway, magnificently restored, in 2007.

ABOVE No 63395 running on the Great Central Railway

The Somerset & Dorset Joint Railway

7F 2-8-0 [1914]

In retrospect it seems curious that having designed a successful 2-8-0 heavy freight locomotive for use on their jointly owned company, the Somerset & Dorset, the Midland didn't adapt it for use on their own railway and relieve their dependence on small 0-6-0 freight engines.

BELOW
Engine 53809

The Somerset & Dorset Joint Railway was operated jointly by the Midland and London & South Western Railways under an Agreement dating back to 1875. Under the Agreement the Midland Railway was tasked with the provision of motive power for the joint railway.

By the early twentieth century the S&D's important coal traffic had grown to one-and-quarter million tons a year. The standard small MR freight engines, adequate for the gentle grades between Derby, Nottingham and London, were proving increasingly incapable of handling the severe gradients between Bath and Bournemouth without resorting to double-heading and banking.

After much pressure from the S&D, Henry Fowler, CME of the Midland, relented and instructed his Chief Draughtsman, James Clayton, to come up with a locomotive design that, unaided, could haul heavy coal trains over the Mendips.

Clayton, given a free-hand, took the opportunity to break with the Midland's "small engine" policy. Although the outside cylinders, Walschaerts valve gear and wheel arrangement were a radical departure, design of the engine was straightforward as it relied mainly on standard components, including the boiler and tender.

The first engine was delivered from Derby in March 1914 followed by five more in August. Immediately successful on the S&D, some testing was undertaken on the Midland mainline with apparently desultory results. However the LMS was happy to add another five engines in 1925, bringing the class total to 11 engines.

Two of the 1925 engines have survived. No 88, which can be seen on the West Somerset Railway, and No 89 which is kept at the Midland Railway Centre, Butterley.

ABOVE Preserved steam locomotive 7F 2-8-0 53809 on the Midland Railway Butterley

South Eastern & Chatham Railway

2-6-0 U CLASS [1917]

RIGHT TOP
SECR K Class 2-6-4T prototype No 790

BELOW
Preserved SECR U Class No 1638

The U and N Class moguls were largely identical two cylindered engines differentiated only by driving wheel size. Three cylinder versions of each class were also built, designated N1 and U1.

Richard Maunsell was appointed CME of the SECR in 1913 and it was immediately clear to him that the railway needed a modern mixed traffic locomotive like the Great Western's 43xx mogul introduced in 1911.

Maunsell recruited a new team which included Harold Holcroft, a Churchward protégé from Swindon, and James Clayton, formerly the Midland Railway's Chief Draughtsman.

They started work on a new 2-6-0 in 1914 but the advent of war meant it was three years before the first prototype, N Class No, 810 was completed at Ashford in 1917. The class proved reliable and competent and continued to be built up to 1934, by which time the class totalled 80 engines.

Developed at the same time and borrowing heavily from the N Class design, was the K Class 2-6-4 express passenger tank designed for the run from London to Dover and with 6 ft driving wheels compared to 5 ft 6 in of the N. The first engine, No 790 River Avon, left Ashford in 1917 although there was then a production delay until 1925 when the next 20 engines, Nos A 791 – A809, were delivered.

One evening in August 1927 River tank No A800 River Cray derailed at speed at Sevenoaks and smashed into a bridge causing 13 fatalities. Acknowledging that one reason for the crash was water surge in the side tanks, it was decided to rebuild all the class as 2-6-0 tender engines and incorporate them into the newly planned U Class moguls.

ABOVE Ropley Station on the Mid-Hants Railway

New U Class engines were built up to 1931 by which time the class, including rebuilds, totalled 50 engines. The rebuildscould be distinguished from the new engines by their lower footplate and slightly larger wheel splashers.

Four U Class Moguls, all of them rescued from Woodham Brothers Barry scrapyard, have survived into preservation. All four have been preserved, appropriately, on Southern heritage sites: two on the Bluebell Railway and two on the Mid-Hants Railway.

24

The Great Northern Railway

N2 0-6-2 TANK [1920]

Nigel Gresley successfully developed Henry Ivatt's N1 0-6-2 tank to produce a sturdy and reliable engine, the N2. All 107 engines of the class passed into British Railway's ownership with the last survivors not being retired until 1962.

During the latter half of the nineteenth century suburban services out of Kings Cross expanded greatly as farms and fields were replaced by housing. For locomotives the route was challenging, with a steady climb over the first 12 miles to Potters Bar and by the early 1900s there was a need to replace the increasingly underpowered 0-4-4 and 4-4-2 tank engines.

CME Henry Ivatt experimented briefly with 0-8-2 tank engines (the L1 Class of 1903) before settling on the 0-6-2 tank as ideal for suburban services in 1906.

The N1s successfully managed the suburban services until after World War 1 when, again, increased loadings demanded something more powerful. By now Nigel Gresley had taken over from Ivatt and after considering a number of tank engine options settled on building an improved NI, the N2.

With larger cylinders, piston valves and superheater the new engines were more powerful than the N1s and the first 60 were delivered during 1920 and 1921. With larger side tanks, a higher pitched boiler and cut down boiler mountings the N2s presented an altogether more imposing appearance than the N1s.

A further 47 of these versatile engines

were built by the LNER between 1925 and 1929. The new engines were allocated to Kings Cross, and were also deployed on similar services round Glasgow, Edinburgh and Dundee.

Some of the engines, like the pictured preserved engine, were fitted with condensing apparatus for underground working on the Metropolitan Railway Widened Lines between Moorgate and Kings Cross. One engine, No 4744 (BR No 69523), has been preserved and it can normally be seen working on the Great Central Railway.

ABOVE
No 1744 at Loughborough on the Great Central Railway

25

The London & South Western Railway

S15 4-6-0 GOODS ENGINE [1920]

RIGHT TOP
SR 828

RIGHT BOTTOM
Southern 825 approaching Levisham Station

Richard Maunsell improved Robert Urie's original somewhat flawed design to produce one of the Southern Railway's finest freight engines, justly deserving their nickname "Goods Arthurs" after Maunsell's "King Arthur" express passenger engines.

Between 1906 until his death in 1912, Dugald Drummond experimented with a number of four cylindered 4-6-0s for the LSWR. They were all relative failures sharing one fundamental flaw: for their size they were too lightly constructed.

Eastleigh's Works Manager, Robert Urie, appointed as CME in 1912 would have been very familiar with the failings of Drummond's engines and built his first 4-6-0, the H15, in 1914: a robust, two cylinder engine, built like a battleship by comparison with his predecessors 4-6-0 designs.

Britain's first truly mixed traffic engines, the H15s quickly demonstrated their success on freight and parcels trains and provided the impetus for Urie's next two 4-6-0 designs, the N15, a passenger express with 6 ft 7 in wheels and the SI5, intended for freight traffic, with smaller 5 ft 7in wheels. Twenty S15s were in service by 1921.

After grouping and with the Southern Railway hungry for new engines, its CME Richard Maunsell, placed an order for a further 15 S15s but only after comparative testing resulted in a radical revision of the front end layout, increase in boiler pressure and modifications to cab and boiler mountings. The changes transferred an adequate engine into a superb freight

engine with a turn of speed that made them equally suitable for passenger expresses.

A further 10 engines were added between 1931 and 1936 bringing the class total to 45 engines, all passing into British Railway's ownership with the last example working until 1966.

Thanks again to the Woodham Bros, seven S15s have survived, each one rescued from their Barry scrapyard. The survivors are spread around the Mid-Hants Railway, the Bluebell Railway and the North Yorkshire Moors Railway, although at present only one is operational, No 30825 on the North Yorkshire Moors Railway.

The London & North Western Railway

0-8-0 G2 [1921]

The final versions of the LNWR's 0-8-0 freight engines, the G1s and G2s were known to the Traffic Department as "D Superheated", quickly abbreviated to "Super D" by loco crews and railway enthusiasts.

BELOW
49395 Super D at Heywood

The London and North Western Railway was one of the first British companies to introduce an eight-coupled freight engine when number 2524 was out-shopped from Crewe in 1892.

Designed by Francis Webb, the LNWR's long serving CME (from 1871 to 1903) and a keen advocate of compounding, number 2524 was quickly followed in 1893 by the first A class compound 0-8-0.

Webb's commitment to compounding meant that by 1900 110 A class three cylinder compounds had been built followed by a further 170 four cylinder compounds built between 1901 and Webb's retirement in 1903.

Webb's successor, George Whale, was vehemently opposed to compounding

and starting in 1904 all Webb's compound engines were rebuilt as simples. Initially designated Class C later rebuilds and new builds with superheaters, bigger boilers and Belpaire fireboxes were classified as Gs and G1s.

The final flowering of the LNWR's 0-8-0s was the G2, credited to Charles Bowen-Cooke but not introduced until 1921, a year after his death.

The main differences between the G2, pictured here, and the G1 were an increase in boiler pressure from 160 psi to 175 psi and the replacement of the engine's vacuum brake with a steam brake. And in appearance there was very little to distinguish the G2 from the first LNWR 0-8-0 built almost 30 years earlier.

Only one G2 (Super D) has survived into preservation; BR number 49395, part of the national collection withdrawn in 1961 and restored to full working order by the Pete Waterman Trust in 2005 at a total cost of £500,000.

ABOVE 5043 Earl of Mount Edgcumbe, Tyseley

27

Great Northern Railway

FLYING SCOTSMAN [1923]

Arguably the world's most famous engine, Flying Scotsman has now worked longer in steam heritage, in Britain and overseas, than the 40 years it spent as a main line express prior to withdrawal in 1963.

BELOW
Flying Scotsman at Steamtown Railway Museum

Flying Scotsman, No 1472, was the third A1 class engine but the first to be delivered as a London & North Eastern Pacific following the railway grouping of 1923.

Nigel Gresley, CME of the Great Northern Railway since 1911, was aware of the need for new engines capable of hauling 600 tons and more but World War I deferred his plans until, in 1922, the first A1 4-6-2 Pacific (No 1470) was delivered from Doncaster works. Gresley was keen to retain the wide fireboxes which had been fundamental to the success of the GNR Atlantics and as this required a trailing truck he opted for the Pacific wheel arrangement rather than the 4-6-0 then popular with other railway companies.

Since 1862 the 10:00am departure from Kings Cross to Edinburgh was named

the Flying Scotsman, probably Britain's most famous named train. In 1924 Pacific No 1472 was specially prepared for display at the British Empire Exhibition at Wembley and named Flying Scotsman.

Exhibited alongside Flying Scotsman at Wembley was the Great Western 4-6-0 Castle Class Caerphilly Castle. To the chagrin of the LNER, subsequent trials between the AI Pacifics and Castle Class showed that the Castles were the more powerful and efficient performers. The tests proved instructive, however, in making informed adjustments to the Pacific's valve gear and boilers. The improved engines were re-classified from A1 to A3 and in 1934 A3 Pacific Flying Scotsman added to its fame when it became the first steam engine to achieve a properly authenticated speed of 100mph.

The sheer expense of keeping Flying Scotsman running has resulted in the bankruptcy of a number of owners, starting with Alan Pegler, who bought the engine in 1963 for £3,000, until in 2004 and with the help of Sir Richard Branson and a grant from the National Heritage Memorial Fund, the engine was acquired for the nation.

ABOVE
Flying Scotsman in Doncaster Works

28

Great Western Railway

4-6-0 CASTLE CLASS [1923]

Although in most respects just an enlarged version of Churchward's Star class of 1907, the Castles proved such reliable and powerful performers that they continued to be built with little modification right up to 1950, by which time the class numbered 171 engines.

With the need for more powerful engines on the West of England mainline CME Charles Collett made a slight increase in size to Churchward's four cylinder Star class 4-6-0s of 1907. A bigger boiler, increased cylinder diameter and enlarged firebox provided a 14% boost in power over the Stars and a recorded tractive effort which established the Castles as the most powerful engines of their day.

The first engine of the class, Caerphilly Castle was delivered from Swindon in August 1923 and promptly put on display at Paddington Station as "the most powerful passenger express passenger engine in Great Britain".

Bearing the same legend the engine was displayed at the British Empire Exhibition at Wembley in 1924 alongside the LNER's much larger A1 Pacific Flying Scotsman. Eager to challenge the claim the LNER agreed to a series of locomotive exchange trials proposed by the GWR's publicity hungry General Manager, Felix Pole.

The comparative trials took place in the spring of 1925, with engines from each company running on the East Coast mainline and the line from Paddington to Plymouth. As could be expected the Pacifics performed well but no one, especially Gresley, was left in any doubt that

the Castles were the superior engines. But Gresley learned from the trials and made informed adjustments to the boilers and valve gear of the Pacifics, resulting in improved performance and an upgrade from class A1 to A3.

Over the next few years the Castles's set a number of speed records, especially on the Cheltenham Flyer, but their main strength was their reliability and consistent day-to-day running.

Eight engines of the class have survived into preservation, including the first engine, Caerphilly Castle, which is on static display at Swindon Steam Railway Museum.

ABOVE
GWR 4079
Pendennis Castle
at Chester General

29

Midland Railway

4F 0-6-0 [1924]

On appointment as CME in 1909 Henry Fowler had the opportunity to break out of the Midland's "small engine" 0-6-0 freight engine policy but settled, yet again, for that ubiquitous model.

BELOW
No 44422 running on a photographic freight charter

In 1845 the Midland Railway, formed only the year before, took over one of England's earliest railways, the Leicester & Swannington and in doing so acquired what are generally regarded as Britain's first inside-cylinder six coupled engines.

Seemingly impressed by their performance Matthew Kirtley, the Midland Railway's first Locomotive Superintendent, adopted the type as his standard goods engine.

And over the next 70 years Kirtley's successors, Samuel Johnson, Richard Deeley and eventually Henry Fowler retained the 0-6-0 as the Midland's freight engine even though, by the twentieth century, other railways had switched to either 2-6-0 moguls or else eight coupled freight engines.

During the nineteenth century the design changed little and though many of the engines survived into British Railways ownership they all suffered from a particular weakness: inadequate bearing surfaces for the driving axles and connecting rods.

Henry Fowler took over from Richard Deeley in 1909 and quickly identified that the Midland's vast freight traffic required modern motive power. Although briefly considering eight coupled engines, Fowler again settled on an 0-6-0 and maintained the Midland's "small engine" policy, frequently requiring double heading on long freights.

The first two prototype engines were built in 1911 with series production starting in 1917. Improvements over the old 3F's were incorporated, superheaters and a vacuum brake so that the engines could work passenger trains, but the new class perpetuated the old problem of inadequate bearings.

In spite of the engine's manifest shortcomings the Midland and its successor, the LMS, built no fewer than 772 4Fs between 1917 and 1941, one of Britain's largest class of engine.

Three LMS built 4Fs have survived, including the first one built, No 44027, as well as a single Midland Railway example, No 43934, which also has the distinction of being the first locomotive to be rescued from Woodham Brothers Barry scrapyard.

ABOVE View SE, LMS 4F 0-6-0 No. 44444 is shunting just south of Stockport (Edgeley) station, 1950. No other British Rail locomotive had a number with five identical digits

30

London Midland & Scottish Railway

0-6-0T JINTY [1924]

BELOW
LMS 0-6-0T 3F locomotive, 16564 (CJ Allen, Steel Highway, 1928)

Mundane work-a-day six coupled tank engines provide little opportunity for development and innovation and though not introduced until 1923 the design of the Jinty can be traced back to a Samuel Johnson design of 1874.

Between 1874 and 1902 Samuel Johnson supervised the building of four similar classes of six coupled tank engines designed for shunting, freight transfer and occasional passenger duties, in total 350 engines.

The last 60 built in 1902, designated class 2441, were characterised by their prototypical round-top fireboxes and dome mounted Salter safety valves. Prior to grouping these engines were rebuilt with Belpaire fireboxes, Ramsbottom safety valves and improved cabs. These rebuilt engines provided the basis for a new standard class of 0-6-0 tank engine for the newly formed London Midland and Scottish Railway. George Hughes, the LMS's new locomotive chief settled on this long established design because it was proven, simple in construction and

straightforward to maintain. A few modifications, increase in cylinder size, boiler pressure, grate area and an extended smokebox, justified an increase in the new engines' power rating from 1F to 3F.

Already fitted with steam and handbrakes, the versatility of the 3Fs was improved by also fitting most of them with vacuum brakes. Extending their availability for passenger work many were fitted with steam-heating and seven were push-pull fitted for working the line between Swansea and Brynamman.

As a testament to the longlived, simple and robust design Hughes' successor, Henry Fowler, continued building the Jinties up to 1931 by which time the class totalled 422 engines.

Ten Jinties survived into preservation with the four currently operational occasionally masquerading as Thomas the Tank Engine.

Just one of the old Midland 1F "halfcabs" survives, BR No 41708, originally built as class 1377 in 1880.

ABOVE A Class 3F "Jinty" locomotive hauled train at Rawtenstall station on the East Lancashire Railway

31

Great Western Railway

56XX 0-6-2 TANK [1924]

RIGHT TOP
GWR 6619
in BR Black

BELOW
5637 at the East
Somerset Railway

When new engines were needed for the steeply graded south Wales lines Charles Collett turned to the 0-6-2tank arrangement previously adopted almost universally by the region's pre-grouping railway companies.

On grouping in 1923 the Great Western took over a number of south Wales railway companies, Taff Vale, Rhymney and Barry, and inherited a number of 0-6-2 tank locomotives favoured by those companies for short-distance dock and colliery work.

The 0-6-2 tank was particularly suited to both forward and bunker first working on the steep sinuous valleys of south Wales so when more engines were needed to meet the 1920s burgeoning coal industry CME Charles Collett opted for a new design of inside cylindered 0-6-2 tank, the 56xx.

Using as many Swindon standard parts as possible 200 engines of the class were built between 1924 and 1928 with the majority being allocated to depots in

South Wales, the region for which they were designed.

Free steaming and powerful the new engines were capable of handling heavier loads than their pre-grouping predecessors though on passenger turns some enginemen expressed a preference for the older engines.

All the engines survived into British Railway's days but withdrawals took place rapidly between 1962 and 1965, New Year's Eve 1965 being the date when steam officially ended on the Western Region.

Nine of the class have survived into preservation, eight having being rescued from Woodhams' scrapyard in Barry. Four are currently in service, four are being restored and one is on static display at the Didcot Railway Centre.

ABOVE 5600 class 5643 arriving at Berwyn on the Llangollen railway

32

The Southern Railway

N15 4-6-0 KING ARTHURS [1925]

The public relations industry was beginning to get established in the 1920's and the decision to name the N15s after King Arthur's Knights of the Round Table was the inspired idea of John Elliot, the Southern Railways advertising and publicity chief.

BELOW SR King Arthur class, 806 Sir Galeron, fitted with wind deflectors for lifting exhaust clear of the Cab. 1928

The man appointed as the CME of the newly formed Southern Railway in 1923 was Richard Maunsell, previously CME of the South Eastern & Chatham Railway, one of the Southern's constituent companies.

World War I and extensive investment in electrification by the Brighton and South Western companies meant the Southern had inherited a relatively aged locomotive fleet. There was nothing capable of meeting the immediate demands of Chief Operating Manager, E.C.Cox, for an express passenger engine capable of hauling 500 ton passenger trains at an average speed of 55mph.

Denied the time and luxury of developing a new passenger express engine, Maunsell settled on the Urie N15 4-6-0s inherited

from the LSWR as a potentially quick solution to the problem.

Developed from Urie's successful mixed traffic 4-6-0s, the H15s, 20 N15s had been built between 1917 and 1923. On paper these should have been excellent express locomotives but in practice they were modest performers with a proven reputation for poor steaming.

Maunsell had faith in Urie's basic design and during 1924 undertook tests with N15 No 742 to try and rectify the fundamentally flawed steam circuit design. The subsequent modifications, outside steampipes, larger steam passages, longer valve travel, redesigned blastpipe and chimney and an increase in boiler pressure to 200psi, transformed the engine. Now it was able to meet the demands of Operating Manager, Mr Cox.

Rebuilding the remaining N15s and some older Drummond 4-6-0's together with new engines ordered from the North British Locomotive Company meant that by 1929 the class had been expanded to 74 engines.

Only one of the 74 engines has survived, No 30777 Sir Lamiel, which was saved as part of the National Collection following withdrawal in October 1961.

ABOVE Restored locomotive Sir Lamiel near Redmire railway station

33

The Southern Railway
4-6-0 LORD NELSON [1926]

Performance from the 16 engines of the Lord Nelson class tended to be disappointing until Maunsell's successor, Oliver Bullied made some radical modifications to the engines' blastpipes.

BELOW Southern Railway (UK) 'Lord Nelson' class 4-6-0 No. 815 Lord Nelson, at Minehead station

In 1923 one of the early demands of the Southern Railway's Traffic Manager, E C Cox, was for express passenger locomotives that could haul trains of 500 tons at average start-to-stop speeds of 55mph.

The Southern Railway had no engines to match this demand but inspired by the GWR's Castle Class the new CME, Richard Maunsell, had begun to think about a powerful four cylinder six coupled passenger express engine that would easily meet Mr Cox's requirement. Redesign of Urie's N15 class into the very successful King Arthurs deferred development of the new engine but in 1925 an order for a prototype, No 850 Lord Nelson, was placed with Eastleigh.

Design of the engine was a compromise between power and weight with

a maximum permitted axle load of 21 tons. Production proceeded slowly as all sorts of ingenious ideas were devised to keep the weight down: thinner than normal mainframes, lighter motion and the use of high tensile steel for the coupling and connecting rods.

Lord Nelson was a handsome looking engine and on paper was for a time Britain's most powerful passenger engine but proved an inconsistent performer over two years of extensive testing. The performance was however considered sufficiently adequate for production to go ahead and an initial order for 30 engines, subsequently halved to 15, was placed with Eastleigh in 1928.

Getting the best performance from a Nelson required skilled firing and footplate crew familiar with the engine, something hard to achieve with a class of only 16 engines. Maunsell continued tinkering with design to improve performance but it wasn't until Oliver Bullied, Maunsell's successor, fitted a five-jet Le Maitre multiple blastpipe and a large diameter chimney that the engines became reliable performers.

Lord Nelson, the doyenne of the class, was preserved as part of the National Collection and is currently in working order and on loan to the Mid-Hants Railway.

ABOVE No 850 at Eastleigh

34

London Midland & Scottish Railway

6P 4-6-0 ROYAL SCOT CLASS [1927]

Born out of hurried necessity and without any prototype testing, the prospects of the LMS's new 1927 4-6-0 class being a success looked unlikely. But following the generous loan from the Southern Railway of the blueprints for their 4-6-0 Lord Nelson class and relying on detailed design work by the North British Locomotive Company, builders of the first 50 engines, expectations were contradicted and the class proved to be competent and reliable performers.

BELOW 46115 - 'Scots Guardsman' hauling the 'Scarborough Spar Express'

During the publicity hungry 1920s the LMS's locomotive fleet, defined still by the old Midland Railway's "small engine policy", compared badly with the Great Western's Castle class, Nigel Gresley's pacifics on the London & North Eastern Railway and the Southern's Lord Nelson class, then Britain's most powerful 4-6-0.

Compounding the problem was the LMS's commitment to accelerate London to Scotland services from the summer of 1927 with timings that could compete with the LNER's east coast timetable. The quickest and easiest solution, to purchase

50 Castle class locomotives, was politely declined by the GWR along with the request to borrow detailed drawings of the class. The Southern Railway responded to requests more graciously and provided detailed drawings of their newly introduced 4-6-0 four cylinder Lord Nelson class.

ABOVE LMSR No. 6100 Royal Scot at Llangollen

Reviewing these drawings and surveying the contemporary locomotives of their rivals, the LMS management concluded they needed a powerful three cylinder 4-6-0 locomotive. Neither Crewe nor Derby had the resources to build the engines in time so an outside builder, the North British Locomotive Company, was contracted to build the first 50 at a price of £7,225 each. The NBL undertook much of the detailed design work, working in conjunction with Herbert Chambers, Chief Draughtsman at Derby, bypassing Henry Fowler, CME of the LMS. Regarded as a competent engineer and metallurgist, Fowler did little during his two decades at both the MR and LMS to establish a reputation as a locomotive designer.

The NBL delivered its 50 locomotives between July and November 1927 and the first engine of the class was named Royal Scot, after the long established named train departing Euston every day at 10:00am for Glasgow and Edinburgh. Sufficiently confident in the new engines' ability, the LMS retimed the Royal Scot to run non-stop over the 299 miles to Carlisle from the 27th September 1927.

Construction continued until 1930 by which time the class totalled 70 engines, all of which were withdrawn between 1962 and 1965. Two engines have survived into preservation, 6100 Royal Scot and 6115 Scots Guardsman, the engine that was featured in the 1936 documentary film Night Mail.

Great Western Railway

KING CLASS 4-6-0 [1927]

The GWR's publicity hungry General Manager, Sir Felix Pole, was happy to send the first King class locomotive, King George V, to the USA to take part in the "Fair of the Iron Horse", celebrating the centenary of the Baltimore and Ohio Railroad. On return she sported a brass bell presented by the Baltimore and Ohio which the engine wore throughout its working life and can still be seen on the preserved locomotive.

BELOW King Edward II at Dereham, Mid Norfolk Railway

During the 1920s there was great rivalry between the "big four" to claim ownership of Britain's most powerful locomotive. The Great Western's General Manager, Sir Felix Pole started the competition when Caerphilly Castle, exhibited at British Empire Exhibition, Wembley, in 1924 carried a notice claiming it to be "Britain's most powerful passenger locomotive".

Stung by the claim the LNER, who were exhibiting A1 Pacific Flying Scotsman at the same exhibition agreed to exchange trials between the A1 Pacifics and the Great Western Castle Class. These took place during 1925 and the Castles were the clear winners.

Having won the duel the Castles were able to retain the crown as Britain's most powerful locomotive until the introduction of Richard Maunsell's 4-6-0 Lord Nelson class on the Southern Railway in 1927. With a tractive effort of 33,500lbs the Lord Nelsons were now, on paper at least, Britain's most powerful locomotive.

Although the commercial justification for building the King Class 4-6-0s was the traffic department's need to run longer and heavier trains there was a suspicion that the General Manager, Sir Felix Pole, wanted to recapture for the GWR the title "Britain's most powerful locomotive". It's certainly true that in a design borrowing heavily from its 4 cylinder Star and Castle predecessors the designer, Charles Collet, was required by Pole to come up with an engine that would exceed 40,000lbs of tractive effort. Careful attention to boiler pressure, cylinder size and wheel diameter meant this was achieved but the axle loading of 22.5 tons limited the engines' route availability and only 30 Kings were built compared to 171 of the rather more flexible Castles.

From new and with only minor modifications to improve their ride the Kings proved themselves admirable performers, reducing the journey time from Paddington to Plymouth to four hours and capable of hauling unaided train loads of up to 500 tons.

Three engines have been preserved, the doyen of the class, No 6000 King George V, 6023, King Edward II and 6024, King Edward I.

ABOVE 6010 King Charles I leaving Paddington, February 1961

36

Great Western Railway

THE HALL CLASS 4-6-0S [1928]

BELOW GWR 4900 Class 4953 Pitchford Hall

The introduction of the Halls in 1928 was an important milestone in the development of mixed traffic 4-6-0 locomotives. They started a tren which would culminate in a series of very successful engines such as Stanier's Black Fives, Thompson's B1s and British Railways Standard Class 5s.

Churcward's 43xx 2-6-0 mixed traffic engines, first introduced in 1911, proved to be very successful engines with 342 being built by 1932. By the early 1920s, however, it was clear that some mixed traffic duties demanded a more powerful engine.

Rather than building a new engine from scratch CME Charles Collett decided to modify an existing Saint Class 4-6-0 engine and chose No. 2925 as his guinea pig. Renumbered 4900 the rebuilt Saint Martin left Swindon in 1924 with new 6' driving wheels replacing the original 6'8 1/2" wheels, re-aligned cylinders, longer travel valves and a more modern cab with a sidewindow.

No 4900 underwent extensive testing which lasted three years and, finally confident with the results, Collett placed an order for a further 80 engines. The first new engine, No 4901, entered service in 1928 with the last engine from the initial batch of 80 being completed in 1930.

The new engines proved so successful, particularly on the challenging Cornish main line, that an order for a further 172 was quickly placed. In total 259 engines of this class had been built when the last engine, No. 6958, was delivered in 1943, two years after Frederick Hawksworth had replaced Charles Collett as CME.

Underlining the success of the class, Hawksworth turned to the design again to meet the wartime demand for further mixed traffic engines. Hawksworth's engines, known as Modified Halls, were substantially the same as the original Halls but showed differences in construction with greater use of fabrication instead of castings dictated largely by the imposition of wartime economies.

By 1950 71 Modified Halls had been built bringing this class of engines popular with both footplate and maintenance staff to a total of 330 engines.

11 Halls and 7 Modified Halls have survived into preservation although only four are currently operational. Two of the survivors are being cannibalised in the reconstruction of a Saint Class, a County Class and a new Grange, all GWR classes of which no example was preserved.

ABOVE GWR 4900 Class 4965 Rood Ashton Hall

Great Western Railway

57XX 0-6-0 PANNIER TANK [1929]

BELOW Great Western Railway 16xx class No. 1638 at Bodiam station on the Kent & East Sussex Railway

Size and weight meant that the 57xx tanks could go almost anywhere and they were a familiar sight all over the Great Western's railway network.

An archetypal and almost exclusive feature of Great Western locomotives, pannier tanks first appeared on 0-6-0s in 1903. Square topped tanks made sense on the Great Western with its preferred use of Belpaire fireboxes and the space left above the frames allowed relatively unrestricted access to the inside cylinders and motion.

Pannier tanks, instead of saddle or side tanks clearly worked for the GWR and became standard for 0-6-0 shunting and branch line engines from 1910 onwards.

Though important as"go-anywhere" workhorses, general purpose small tank engines were not included in Churchward's scheme of standard loco-motives but by the 1920s it was clear to Churchward's successor, Charles Collett,

that many of the older and rebuilt 0-6-0 tanks were time expired and in need of replacement. He settled on a new standard class of small tank engine which, drawing upon proven and reliable designs, required no prototype and an initial order for 100 was placed in 1929.

Assigned class 57xx, 300 had been built by 1932, 200 funded under a government scheme to tackle unemployment during the depression. Building continued under British Railways and by 1954 there were 863 engines, making it the largest class of engine on the Great Western.

Numbers in the 57xx series were quickly exhausted and later engines used available numbers in the 36xx, 37xx, 46xx, 67xx, 77xx, 87xx, 96xx and 97xx series.

As useful today in steam heritage as they were in their working heyday, 16 of these tanks have survived into preservation.

ABOVE Auto train fitted Pannier Tank No 6424 heading an auto train from Bewdley in 1963

38

London Midland & Scottish Railway

7P 4-6-2 PRINCESS ROYAL CLASS [1933]

RIGHT TOP Princess Margaret Rose at Swanwick on the Midland Railway Butterley heritage site

RIGHT BOTTOM 6201 Princess Elizabeth Castleton East Junction

When William Stanier arrived as the LMS's new CME at the start of 1932 one of his first priorities was the design of an engine that could haul 500 ton trains unaided on the 401 mile run from Euston to Glasgow. The Royal Scot 4-6-0s, introduced to the service in 1927, were putting in respectable timings but were limited to engine loads of 420 tons and needed an engine change at Carlisle.

The first prototype, Pacific No 6200, Princess Royal, was delivered from Crewe in June 1933 and displayed clearly the influence of Stanier's 30 year career on the Great Western, the last 10 as Charles Collett's principal assistant. Prime influence for the new engine was the Great Western's King class, with which it shared cylinder dimensions, boiler pressure and couple wheel size. The main departure, requiring a Pacific rather than 4-6-0 wheel arrangement, was a massive boiler and firebox requiring the support of the Pacific's trailing truck.

Though 6200, in its gleaming crimson lake livery, looked impressive, performance didn't match. With its restricted steam circuits and low degree superheat the engine was a poor steamer. But with the problems identified and adjustments made to 6200 and the second prototype, No 6201 Princess Elizabeth, delivered in November 1933 the engines' performance began to live up to their appearance.

Stanier continued tinkering with the two prototypes until 1935 by which time he was sufficiently confident to confirm production of 10 further engines, with the first of the batch, No 6203 Princess

Margaret Rose, named after Princess Elizabeth's younger sister.

During 1937 two test runs by Princess Elizabeth, the second well-advertised, established the class as record breaking performers. That second test, undertaken on the 16th and 17th of November 1937, knocked 16 minutes off the six hour Glasgow to Euston schedule and won Princess Elizabeth the public's affection. The class of 12 engines was officially the Princess Royal Class but in recognition of the public's endearment they were universally referred to, by train-crew and enthusiasts alike, as "Lizzies".

All 12 of the class were withdrawn between 1961 and 1962 but two engines have been preserved in working order, "Lizzie" herself and No 6203, Princess Margaret Rose.

39

London Midland & Scottish Railway

MIXED TRAFFIC 4-6-0 "BLACK FIVE" [1934]

BELOW 44949 at Manchester Victoria, 1968

Drawing on his detailed knowledge of the successful GWR mixed traffic Hall class, William Stanier was able to design an improved mixed traffic engine, the "Black Fives", for the LMS. The design was so successful that the class totalled 842 engines when production ceased in 1951.

In the 1920s and with almost 8,000 varied route miles, the LMS had acknowledged the need for a new general purpose mixed traffic engine.

When William Stanier joined the LMS as CME in 1932 he had worked at Swindon for 35 years and was totally familiar with the Great Western Railway's very successful 2 cylinder mixed traffic 4-6-0, The Hall class.

Starting with his knowledge of the Halls, Stanier improved on it in designing the new 5P/5F utility design for the LMS. The cylinder casting with extended valve chests provided freer steaming and a higher top speed than the Halls. With the LMS '5' power rating and a lined black livery the name "Black Five" was quickly coined for the new engines.

The first 50 "Black Fives" were built straight from the drawing board without the luxury of prototype testing although Stanier had been able to experiment with some of his ideas on his first new engines for the LMS, the 40 strong class of moguls built in 1933, just ahead of the Black Fives. Crewe provided the first 20 engines with 30 commissioned from Lancashire locomotive builder Vulcan Foundry. Vulcan delivered their first engine, 5020, six months ahead of the first engine, 5000, due from Crewe.

Immediate tests of No 5020 on both passenger and freight trains proved it to be a very competent performer and an LMS report concluded that "The 2-cylinder 4-6-0 appears to be a very efficient and satisfactory mixed traffic power unit".

With its proven success the LMS quickly went on to build the "Black Five" in volume and before World War II interrupted production the company had amassed 472 engines, built variously by Crewe and a range of external contractors. Production resumed after the war and by 1951, when building ceased, the class totalled 842 engines, making it one of Britain's largest class of locomotives.

Some members of the class worked right up to the last day of steam on British Railways in August 1968 and a total of 18 have been preserved.

ABOVE Black Five 45379 visiting Oldland Common railway station

The Southern Railway

SCHOOLS CLASS 4-4-0 [1934]

Purpose built for the very exacting challenges of the line between Tunbridge Wells and Hastings the "Schools" quickly established themselves as exceptional performers highly regarded by their train crews.

Cheaply constructed in the 1850s the line from Tunbridge Wells to Hastings had a twisting, turning track and limited clearances, particular in Mountfield Tunnel, near Battle, where the skimped construction of the original tunnel had to be repaired with an extra lining, limiting its width clearance to only 8' 61/2".

This ruled out larger express engines and the line was the preserve of old pre-grouping 4-4-0s hauling narrow bodied coaches constructed specially for the line. By the 1920s Hastings was developing as a popular sea-side resort and London's commuters were starting to move out to Kent and Sussex. With increasingly longer and heavier trains there was an urgent need for more powerful locomotives.

The specification was demanding: cab width could be no more than 8'6", height was limited to 12' 103/4" and existing turntables and the line's sharp curves imposed a shortened frame.

Richard Maunsell and his drawing office staff wrestled with the problem and came up with an inspired solution: the V Class 4-4-0.

The new engines had three cylinders, to ensure both adequate power and outside cylinders small enough to fit the loading gauge, a specially profiled cab, a round

ABOVE Schools Class locomotive No 925, Cheltenham, at National Railway Museum, York

topped firebox to provide adequate forward vision from the narrow cab and incorporation of as many "standard" parts as possible from the Lord Nelsons and King Arthurs.

No prototype was built, the first 10 engines being ordered and built straight from the drawing board in 1930. The engines immediately proved themselves excellent steamers with responsive controls and quickly established themselves as Europe's most powerful 4-4-0s easily capable of handling 350 ton trains between London and Hastings. Their crews loved them.

By 1935 the class was complete at 40 engines all named after public schools, initially within the boundaries of the Southern Railway.

In the late 1950s Kent coast electrification and the construction of narrow bodied diesel-electric multiple units rendered the Schools redundant, the class being fully withdrawn by the end of 1962.

Three engines have survived, No 925, Cheltenham was reserved for the National Collection and two others, 926, Repton and 928 Stowe, have been privately preserved.

41

London Midland & Scottish Railway

JUBILEE CLASS 5XP 4-6-0 [1934]

Even by the 1930s the LMS was still suffering from the overhang of the old Midland Railway's "small engine" policy with a shortage of sufficiently powerful and reliable passenger locomotives. The solution was the design and construction, straight from the drawing board, of a large number of three cylinder 5XP 4-6-0 express engines.

Using elements of Stanier's successful Patriot and Black 5 classes and drawing on his experience of GWR design under the tutelage of G J Churcward there was every reason to expect the engines to perform well without the need for extensive prototype testing.

Trouble was the early engines were dismal failures, with a reputation for poor steaming and excessive coal consumption. Crews on the old Midland section preferred the older and theoretically less powerful 4-4-0 compounds to the new Jubilees. The engines looked like being Stanier's first significant failed design for the LMS.

Much trial and error experimentation with superheaters, boiler tubes and blast pipe dimensions was required to identify, isolate and resolve the problems with the Jubilees – but not before 113 of the unmodified engines were in service.

Continued tinkering and improvement of the engine design up to the Second World War resulted in successful locomotives popular with engine crews, with most of the 191 strong class surviving up to the late 1960s.

When first introduced the Jubilees were

ABOVE No 45593 Kolhapur in British Railway's standard Brunswick gree livery

painted in LMS crimson and referred to as the "Red Staniers", distinguishing them from the Black 5 4-6-0s which were described as "Black Staniers". To mark the Silver Jubilee of King George V and Queen Mary in 1935 engine 5642 was named Silver Jubilee and painted in a black and silver livery. At the same time its number was swapped with 5552, the first engine of the class, and thereafter the engines were known as Jubilees or Jubs.

Four of the class of 191 engines have survived though all are currently undergoing overhaul: No 5593 Kolhapur, No 5596 Bahamas, No 5690 Leander and No 5699 Galatea.

London Midland & Scottish Railway

8F 2-8-00 [1935]

BELOW Stanier 8F No 48476 at Lostock Hall shed a few days before the final end of steam on British Railways in July 1968

To address the company's problem of an aging and underpowered locomotive fleet William Stanier was given the authority to "scrap and build" when he joined the LMS as Chief Mechanical Engineer in 1932. 1933 and 1934 saw construction of Stanier's mixed traffic Black 5s and express passenger Jubilees as well as production plans for a 2-8-0 freight engine, the first 12 of which were delivered in 1935.

The two cylinder, 2-8-0's design was based on the successful Black 5 4-6-0 and the overall concept drew heavily on Stanier's first-hand knowledge of Churcward's GWR 28XX 2-8-0 heavy freight locomotive, a tried and tested design dating back to 1903.

The LMS engine's Belpaire firebox and tapered boiler confirmed its GWR heritage but the cylinders and valves had the identical dimensions to the Black 5, with outside Walschaerts gear instead of the inside Stephenson's valve gear favoured by Churcward.

Based at Willesden, Wellingborough and Toton the first 12 engines proved an immediate success, easily managing 1,000 ton coal and ore trains but also capable of handling summer passenger trains at speeds in excess of 60mph. Though versatile and efficient, the LMS's production remained focused on the Black 5s and by 1939 only 126 of the 8F 2-8-0s had been built.

ABOVE LMS Stanier Class 8F 2-8-0 No 8274

At the outbreak of the Second World War, however, the War Department's Director of Transportation, Robert Riddles, had no hesitation selecting the 8Fs as the most appropriate eight-coupled "engines of war". Over 200 engines were sent to the Middle East and to meet demand at home hundreds of engines were built at the workshops of the LMS, LNER, GWR and SR as well as by third party contractors like North British and Beyer Peacock.

Final construction totalled 852 engines, making it one of Britain's largest locomotive classes, with 666 passing into British Railways ownership on nationalisation in 1948. Significant numbers of Lancashire based engines remained in use until 4th August 1968, the last day of steam on British Railways and seven have survived in preservation. In an odd quirk of history these seven survivors have NOW been joined by three engines returned from Turkey, part of the consignment of War Department engines that had continued in use there until the 1980s.

43

London & North Eastern Railway

A4 4-6-2 PACIFIC [1935]

BELOW LNER Class A4 4498 (BR 60007) Sir Nigel Gresley doing the Forth Circle route approaching Stirling from Alloa

Nothing better epitomised the so called Golden Age of steam than the LNER's streamlined A4 locomotives popularly referred to as "streaks", the culmination of Nigel Gresley's three cylinder Pacific design which had commenced with A1 Pacific No 1470 Great Northern, delivered to the Great Northern Railway in 1922. The third A1 Pacific and the first to be delivered to the newly formed London and North Eastern Railway was No 1472 Flying Scotsman, arguably the world's most famous steam locomotive.

Significant improvements to the A1's valve gear and boilers, informed partly by comparative locomotive tests between Flying Scotsman and the Great Western's Caerphilly Castle, saw the launch in 1928 of a new class A3 "Super Pacific". Subsequently all but one of the 52 A1 Pacifics were rebuilt as A3s.

With the constant rivalry for accelerated timings between the LNER and LMS, Nigel Gresley, inspired by the streamlined diesel-electric railcar Fliegende Hamburger on German State Railways, thought that streamlining an A3 Pacific might give the LNER the edge on locomotive speeds.

After confirming that A3 Pacifics could readily sustain speeds of over 100mph and following extensive testing of Plasticine models in the National Physics Laboratory's wind tunnel, the first streamlined A4 Pacific, No 2509 Silver Link was delivered in September 1935.

Like Silver Link, the next three A4s included the word Silver in their names and were used to inaugurate the London to Newcastle Silver Jubilee serviced, named in recognition of King George V's Silver Jubilee.

Designed for fast running, the A4s were soon setting new timing and speed records, culminating in the legendary run of No 4468 Mallard on the 3rd July 1938 when it set the world steam speed record of 126mph, a record unchallenged to this day.

In 1937 the LNER board agree to the proposal of rail enthusiast K.Risdon Prentice that Gresley's 100th Pacific, Doncaster works number 1863, should be named after the designer. The locomotive, No 4498 Sir Nigel Gresley, entered service on 23rd November 1937 and was formally named three days later at Marylebone Station with Sir Nigel Gresley in attendance.

The A4s remained on top-link duties until the early 1960s when their east coast duties were taken over by Deltic diesels. A few of the engines were transferred to Scotland where the final survivors continued working up to 1966.

Of the total class of 35 engines six have been preserved, including BR No 60007 Sir Nigel Gresley which is currently running in the handsome but short lived British Railways experimental blue livery illustrated here.

ABOVE A4 No 60009 Union of South Africa

44

London & North Eastern Railway

V2 2-6-2 GREEN ARROW [1936]

RIGHT LNER Class V2, 60800 "Green Arrow" at Crewe Works open day on 1st June 2003. This locomotive is the sole survivor of its class, and is seen in its British Railways guise.

In the mid 1930s the LNER offered a fast freight service with guaranteed next-day delivery branded Green Arrow. Green Arrow was the name bestowed on the doyenne of the V2 2-6-2 class of mixed traffic engines designed especially for express passenger and fitted freight work.

In 1934 the LNER tasked its Doncaster drawing office to come up with a new design for a mixed traffic engine capable of "heavy, long distance work".

A number of wheel arrangements were considered but Doncaster finally elected for a 2-6-2 Prairie type with a trailing truck that would support a wide firebox and provide better riding at speed than the K3 moguls.

The first five engines, including 4771, Green Arrow, were delivered in November 1936 and then dispersed to London, Peterborough, York and Dundee for extensive testing. Trials showed the engines to be free-steaming, free-running and capable of working vacuum-braked freights at 60mph.

In addition the engines could readily deputise for Gresley's Pacifics which they easily matched in sustained high speed running. Doncaster shed regularly put a V2 on the down Yorkshire Pullman which was scheduled to reach King Cross in "even time", 156 miles in 156 minutes.

Once proven as versatile performers orders for new engines were placed and by 1944 the class totalled 184 engines.

The class survived until near the end of

steam, the first withdrawals starting in 1962 with the last engine, No. 60831, being retired on the last day of 1966.

As Britain's premier class of powerful mixed traffic engine and the only Prairie tender engine to be produced in high numbers, Green Arrow was selected for preservation by the British Transport Commission.

Altough originally conceived as a static exhibit Green Arrow has beeb restored to running order a number of times since withdrawn in 1962 and remained a popular main-line performer until expiry of its boiler ticket in 2008.

45

London Midland & Scottish Railway

8P 4-6-2 DUCHESS PACIFIC [1937]

BELOW LMS Coronation Class Pacific No 46229 Duchess of Hamilton at Birmingham Railway Museum

The so called Golden Age of Steam is best exemplified by each of the big four railway companies most powerful passenger locomotives: the Great Western's King class, Southern Railway Lord Nelsons, London and North Eastern Railway's A4 Pacifics and on the London Midland and Scottish the Duchess Pacifics. Being able to boast the fastest or most powerful locomotive was important for the railways' public relations departments and sometimes it seemed as though the justification for each new more powerful design was simply to trump the prowess of rivals' locomotives.

The impetus for the design of the Duchess Pacific's was the LNER's announcement of its new streamlined Coronation service, named in honour of the King's 1937 coronation, and scheduled to complete the journey between Kings Cross and Edinburgh in just six hours.

This was a challenge the LMS could not afford to ignore but it could only muster 12 engines, the Princess Royal Pacifics, to match the speed, timings and frequency provided by the 80 Gresley Pacifics of the LNER. The LMS quickly needed new and more powerful engines.

Stanier's Princess Royal class provided the starting point for the new design but as Stanier had been seconded to the Indian State Railways by the British government, detailed design work was done

ABOVE Coronation Class Pacific No 6229 on display at the National Railway Museum, York, in LMS livery and with its pre-war streamlined casing restored

by his capable assistant, Robert Riddles and chief draughtsman, Tom Coleman.

The main changes to the Princess Royal design were a larger and much improved boiler, replacement of the complicated four sets of valve gear by two sets of outside Walschaerts motion operating the inside cylinder valves through rocking shafts and an increase in the driving wheel diameter by 3" to 6' 9".

Although the LMS's detailed investigations since 1931 had confirmed locomotive streamlining was of negligible benefit the first 10 locomotives were built with streamlining, really for publicity reasons and to underline the competition with the LNER's A4 "streaks". The remaining 28 engines of the 38 strong class were built without streamlining, as illustrated here, and from 1946 onwards streamlining was removed from all the earlier engines as they passed through the works.

Officially styled Princess Coronation Pacifics but always popularly referred to as Duchesses after the names bestowed on the second set of five to be built, the engines soon established speed, timing and power output records, demonstrating their considerable advance over the Princess Royal design.

With a tractive effort of 40,000 lb ft the Duchesses were Britain's most powerful passenger steam locomotives, a record never overtaken: even by British Railways single 8P Pacific, The Duke of Gloucester.

Three Duchesses have been preserved, two in working order, No 6229 Duchess of Hamilton and No 6233 Duchess of Sutherland, and one, No 6235 City of Birmingham, as a static exhibit.

46

London & North Eastern Railway

K4 & K1 2-6-0S [1937]

BELOW LNER Class K4 The Great Marquess at Stirling

Gresley's K4s, a class of only 5 engines specially designed for the West Highland Line, provided the basis for the very successful class of Peppercorn designed K1s, 70 engines built between 1947 and 1950.

During the 1930s former Great Northern 2-6-0s and North British 4-4-0s were struggling with increasingly heavy trains on the West Highland line from Glasgow to Fort William and Mallaig. With his penchant for tailoring locomotives to particular situations Gresley set about designing the most powerful three cylinder mogul permissible on the line, and one that would end the need for double-headed trains.

The prototype K4, No. 3441, Loch Long with smaller wheels and boiler than the K3 moguls, left Darlington in 1937 ready for testing on the West Highland line. Modifications after testing established Loch Long as a powerful and economical performer and a further four engines had been added to the class by 1939.

Declining maintenance standards during

World War II meant poorer performance from the K4s with their three cylinders and relatively complicated conjugated valve gear. In 1945 and in an attempt to produce a more robust and simpler standard 2-6-0 Gresley's successor, Edward Thompson, rebuilt K4 No. 3445 MacCailin Moor as a two cylinder engine with boiler pressure increased from 200psi to 225psi.

Year long tests across the extensive North Eastern network proved 3445 to be an efficient, reliable and economic performer. By now Thompson had been succeeded by Arthur Peppercorn who proposed further improvements to the prototype resulting in an order for 70 class K1 engines, from the North British Locomotive Company in 1947. In the careful way of the London and North Eastern, the prototype engine, number 3445, was reclassified as class K1/1 to distinguish from the main class.

ABOVE British Railways built and restored K1 2-6-0 No 62005 seen repainted and renumbered in LNER livery at Steamtown Railway Museum, Carnforth, in February 1983

47

Great Western Railway

4-6-0 MANOR CLASS [1938]

BELOW GWR Manor Class 4-6-0 locomotive 7802 Bradley Manor, having finished taking on water at Bridgnorth railway station

Introduced in 1938, the Manors' were initially regarded as mediocre engines. After 1951 some simple modifications transformed the engines' performance and no less than 9 have survived into preservation on Britain's heritage railways.

At grouping in 1923 the Great Western Railway absorbed a number of minor railways whose track was too lightly laid for the standard range of Great Western tender locomotives.

To meet ongoing traffic demands the engines taken over from companies such as the Cambrian and the Midland & South Western had to be kept running although by the 1930's it was clear that something more modern was needed to replace these increasingly aged engines.

A first low cost attempt to meet this demand was the Dukedog class, a cannibalised rebuild of old Bulldog class frames combined with boilers from the Duke class. Sticking with the economic policy of re-using parts from older engines Charles Collett devised the 78xx Manor Class in 1938, an engine specifically designed with a low axle-loading for use on lightly laid lines.

The new engines used motion and driving wheels from scrapped 43xx moguls paired with second-hand Churchward tenders originally built before World War I. New frames, cylinders and boiler, the Swindon No. 14, were designed for the class, 20 of which had been built by February 1939.

The engines were not a success with a reputation for poor steaming that made them unpopular with footplate crews. In spite of that a further 20 were ordered but World War II interrupted their construction until 1950 when half the order, 10 engines, were built by British Railways.

In an attempt to resolve the steaming deficiencies of the Manors one of the class, No. 7818, Granville Manor, was subjected to an intensive examination by the test team at Swindon in 1951. The team discovered that only minor changes were needed to produce a dramatic change in the engine's performance. A smaller blastpipe tip, longer and narrower chimney with a reduced choke combined with increased air space in the firegrate increased boiler output by an amazing 120%. Over the next few years these modifications were applied to all 30 engines of the class.

ABOVE Carrying a "Going With Regret" headboard, GWR Manor Class 4-6-0 locomotive 7802 Bradley Manor at Kidderminster Town railway station after a southbound Severn Valley Railway journey from Bridgnorth

48

Southern Railway

MERCHANT NAVY CLASS 4-6-2 PACIFIC [1941]

RIGHT TOP
SR Merchant Navy class 35028 Clan Line, west of Bath, on the former-GWR Bristol-Paddington mainline

RIGHT BOTTOM
No 35027 Port Line at work on the Bluebell Railway

On paper and with a reputation for engineering vision Oliver Bullied looked eminently suitable for the role of Chief Mechanical Engineer at the Southern Railway, the post he was appointed to in 1937. Starting with a premium apprenticeship at Doncaster in 1901, married to the daughter of locomotive engineer H A Ivatt and with twenty five years spent as Nigel Gresley's assistant and right-hand man his CV looked unassailable.

Bullied embraced the role wholeheartedly, getting involved in imaginative design of all-steel steam and electric coaching stock and undertaking a detailed assessment of the Southern's entire steam locomotive stock.

From that assessment Bullied was able to persuade Southern management of the need for new modern locomotives and in 1938 won board approval for his new express passenger engine design. But by the time it was ready to go into production Britain was at war and Government policy required that only simple freight and mixed traffic engines could be built.

Bullied sidestepped that diktat by describing his new Merchant Navy class as mixed traffic engines, a fiction that was quickly exposed when the first engine, Channel Packet, was unveiled. Bursting with engineering innovations and imposing looking with its unique air-smoothed casing Channel Packet was a 100% thoroughbred passenger express engine: anything but simple.

Apart from the air-smoothed casing,

which earned the engines the wartime soubriquet "spam-cans", other innovations included thermic syphons, American style "Boxpox" wheels and, importantly, chain driven valve gear housed in an enclosed oil bath.

Well maintained and with their almost limitless steaming capability the Merchant Navies could, when well driven, produce legendary performances. But, sadly, the unique feature of Bullied's Pacifics, the enclosed chain driven valve gear, also proved to be their Achilles Heel. Temperamental and erratic in operation it also seemed impossible to completely seal the oil bath, with leaking oil causing engine slipping, fires in the boiler lagging and all too frequent engine failure.

To resolve these problems the complete class of 30 engines were rebuilt with Walschaerts valve gear and a generally more conventional appearance between 1957 and 1959 as illustrated here by no 35028 Clan Line.

The rebuilt engines were reliable and economical performers, all lasting to the end of steam on the Southern Region with 11 engines surviving into preservation though, sadly, all in the rebuilt form which somehow lacked the magical aura of the original unmodified engines.

London & North Eastern Railway

B1 4-6-0 [1942]

RIGHT TOP
LNER Thompson Class B1 No 61264 piloting the other preserved B1, No 1306 Mayflower, at Barrow Hill Engine Shed in 2007

RIGHT BOTTOM
Class B1 4-6-0 No 1306 Mayflower at a Doncaster Works open day on 27th July 2003. The name, number and apple green LNER livery is not authentic as the engine was built by British Railways after nationalisation

Edward Thompson, Doncaster's mechanical engineer, was not an obvious candidate for the role of CME after Sir Nigel Gresley's untimely death in 1941: more a Hobson's choice forced on a reluctant LNER board. Thompson was 60 at the time of his appointment and though having spent all his working life as an engineer on both the North Eastern Railway and Great Northern Railway had relatively little experience of locomotive design. One of his guiding tenets, though, was a belief in standardisation of locomotive design and production. An idea he was able to exploit practically by producing an engine to meet the LNER'S wartime demand for a mixed traffic 4-6-0.

One of Nigel Gresley's guiding policies was the design of engines to meet specific tasks, bequeathing the LNER with many small classes of engine and no high volume class of mixed traffic engine comparable to the LMS Black 5s or GWR Halls. Gresley's V2 2-6-2 mixed traffic engines were excellent performers, but too heavy for much of the LNER'S system, especially the former Great Eastern lines.

Constrained by wartime limitations on resources and materials, Thompson's solution for the design of a new 4-6-0 was to source components from available and reliable stock designs, using existing patterns, jigs and tools and replacing traditional castings, where possible, with steel fabrications.

So, like building a kit, the Type 100A parallel boiler came from Gresley's B17

Football class 4-6-0s, the 6' 2" wheels from the V2 class, the chimney from the O2 2-8-0 while the tender was adapted from the K3 2-6-0. And for rugged simplicity and to save both weight and maintenance the engine had just two cylinders, adapted from the K2 2-6-0 castings.

Temperamental and unloved by his subordinates, workers and a generation of train spotters, there is a tendency to denigrate Thompson's work as a locomotive engineer. So how did his mongrel 4-6-0, styled class B1, perform? Rather well as it happens and though often compared unfavourably with the Black 5, an engine designed and produced during the luxury of peacetime, the Locomotive Exchange trials of 1948 confirmed the B1 a deserving equal to Stanier's 4-6-0.

As further confirmation of its all round, go anywhere usefulness, construction continued up to 1952 by which time the class numbered 410 engines.

Unlike ex Southern and Great Western engines, only one ex LNER locomotive was sent to Woodham Bros Barry scrapyard. That engine was B1 No 61264 and though rescued in 1973 it wasn't returned to steam until 1997, making it just the second B1 to be preserved.

50

Southern Railway

Q1 0-6-0 [1942]

RIGHT Mr. Bulleid's Ugly Duckling, but a fine engine in service. The long firebox on a 'Charlie' takes a little getting used to but once conquered what a machine this is. 33001 captured at Sheffield Park on a cold winter's day

When unveiled at Charing Cross station in May 1942 the first Q1 attracted a lot of opprobrium because of its unusual and unconventional appearance. William Stanier, when he first saw the engine, reputedly asked "Where do put the key?" But coming from the febrile mind of Oliver Bullied the engine's appearance reflected a precise and timely response to a series of problems.

Passengers rather than freight were the Southern Railway's main source of income, reflected, during the 1920s and 30s, by major investment in electrification with an attendant failure to build new steam locomotives in any significant volume.

With the outbreak of World War II the Southern Railway suddenly assumed logistical importance with its proximity to Europe and the requirement to manage large troop movements and transport vast quantities of supplies demanded by the conflict.

To provide maximum flexibility and access to all the Southern's routes suggested an 0-6-0 locomotive rather than the heavier Stanier 8F 2-8-0 adopted as Britain's official "war" engine and built in considerable volume by the War Department.

The problem, though, was that the newest 0-6-0s the Southern could muster were 20 Q Class locomotives, Richard Maunsell's last design for the Southern Railway delivered after his retirement in 1937. With their design and construc-

tion compromised by the need for simplicity and requirement to run on even the most lightly laid Southern track the Q class were distinctly mediocre performers with a reputation as poor steamers.

In August 1940 Bullied got Southern Railway board approval for the construction of 40 powerful 0-6-0 freight engines. His challenge was to design the most powerful 0-6-0 possible within the Civil Engineers weight constraint of 54 tons for the engine and 391/2 tons for the tender.

Bullied's solution was to concentrate weight where it counted, the engine's power source, its boiler and firebox and then seek compensating weight savings in other areas of the engine's design. Weight savings were achieved by dispensing with running-plates and wheel splashers, replacing conventional castings with fabrications and employing American inspired "Boxpok" wheels with a 10% weight saving compared to traditional spoked wheels.

And the distinctively unique appearance of the boiler was dictated by the use of lightweight fibreglass lagging unable to support the weight of conventional boiler cladding.

But in spite of their ugly appearance, the engines, known universally as Charlies, were a complete success. Weighing in at just over 51 tons and generating tractive effort of 30,000 the locomotives were around 14 tons lighter than any comparably rated freight engines.

The first engine of the class has been preserved as part of the National Collection and can be seen at the National Railway Museum, York.

51

Southern Railway

WEST COUNTRY & BATTLE OF BRITAIN CLASS 4-6-2 LIGHT PACIFIC [1945]

RIGHT SR Bullied "Battle of Britain" Class 4-6-2 No.34067 (ex-SR No.21C167) "Tangmere", although the nameplates are not yet fitted following overhaul in BR Brunswick Green (with 1956 totem) between Bath and Keynsham on the Three Bridges-Bristol Temple Meads "Bath and Bristol Express"

When Oliver Bullied got board approval for construction of the Merchant Navy Pacifics there was also recognition of the need for more new locomotives, but lighter and with greater route availability, to replace the aging steam locomotives in use on the unelectrified West Country and Kent Coast services. But with Brighton Works busy building LMS designed 8F 2-8-0s for the war effort, construction of these new engines had to be deferred until 1945.

During the war years Bullied considered what these new mixed traffic engines should be like and though early drawings suggested a 2-6-0 or 4-6-0 locomotive the final decision was for a smaller and lighter version of the Merchant Navy Pacifics. The main argument for this approach was one of standardisation and technically the two engines were virtually identical, sharing the same 6'2" driving wheels and chain driven valve gear. But with a reduced diameter boiler, shorter frames and a smaller firebox the weight of the new engine was reduced to 86 tons which, together with their axle loading of less than 19 tons, meant they could be used on all the Southern's lines.

Primarily intended for use on the Southern's Western and South Western lines the earlier engines were all given West Country place names, establishing their description as West Country Pacifics. The first engine of the class, Exeter, was appropriately unveiled at Exeter on the 10th July 1945, resplendent in bright malachite green with yellow sunshine lettering and lining. Construction

proceeded at a pace, typically two to four engines a month, and by 1951 the class totalled 110 engines, 40 built under the auspices of British Railways.

As newly built engines started working on the Southern's central and eastern sections a clever public relations ploy saw them named after wartime airfields, RAF squadrons and personalities involved in winning the battle in the skies over Kent, Sussex, Surrey and Hampshire, explaining these engines' designation as Battle of Britain class.

The Light Pacifics suffered the same operational problems as their larger stable mates, the Merchant Navies, and rebuilding in a more conventional form with Walschaerts valve gear started in 1957. When the rebuilding programme stopped in 1961 60 of the class had been modified and 50 were left in their original condition. The weight of the rebuilt engines increased to 90 tons, precluding them from service on lines to Barnstaple, Padstow and Ilfracombe.

Withdrawals started in 1963 and most of the engines that ended up in Woodham Brothers Barry scrapyard were subsequently rescued, explaining the high number of Light Pacifics that have been preserved. Though some are still under restoration and overhaul, 10 rebuilt and 10 un-rebuilt engines have survived.

52

Great Western Railway
10XX COUNTY CLASS 4-6-0 [1945]

BELOW Hawksworth 4-6-0 No. 1011 'County of Chester'

With over 300 Halls and 80 Granges did the Great Western Railway really need another 30 newly designed 4-6-0s? None of the class was preserved but the Great Western Society is currently recreating a new County, No. 1014, County of Glamorgan.

A keen proponent of standard design and standard components, George Jackson Churchward drew up a scheme for six standard Great Western Railway locomotive types in 1901.

These included two 4-6-0 express passenger engine designs: the 2 cylinder "Saints" and the four cylinder "Stars".

Under subsequent CMEs, Charles Collett and Frederick Hawksworth, both classes evolved: the four cylinder Stars into Castles and Kings and the two cylinder Saints into Granges and Counties.

The Counties, introduced in 1945 rep-

ABOVE Hawksworth 4-6-0 No. 1026 'County of Salop'

resented the culmination of Saint design and the 30 engines of the class were apparently built to provide a more powerful alternative to the Halls and Granges.

Adequate but unexciting performers, the first engine of the class, No.1000 County of Middlesex, was subjected to extensive testing in 1954 by Sam Ell, Swindon's acknowledged expert in blastpipes and draughting arrangements. This resulted in all the class being fitted with double blastpipes and new, squatter, double chimneys.

The need for building the Counties was often questioned and withdrawal of all 30 engines by 1964 tends to confirm they were superfluous to requirement. None of the withdrawals ended up in Dai Woodham's Barry Island scrapyard and as a consequence none were preserved.

However the Great Western Society is recreating a new County class engine, No 1014 County of Glamorgan, from the frames of a Hall and the boiler of a Stanier 8F 2-8-0.

53

London Midland & Scottish Railway

CLASS 2 2-6-0 AND 2-6-2T [1946]

RIGHT TOP
LMS Ivatt Class 2MT 2-6-0 no. 46443 at Bewdley

RIGHT BOTTOM
41241 at Haworth

The emphasis on both crew comfort and operating efficiency produced two classes of engines that were particularly well liked by both footplatemen and shed staff.

Following the sudden death of Charles Fairbairn in 1945 H G Ivatt was appointed CME of the LMS in his place.

The ravages of war meant that Ivatt faced a particular problem: the urgent need to replace poorly maintained and time expired pre-grouping engines, particularly 0-6-0s and 2-4-2 suburban tanks.

For the replacement engines Ivatt settled on a 2-6-0 mogul and a near identical 2-6-2 tank engine version. Recognising a general post-war labour shortage Ivatt was concerned that the replacement engines should be operationally efficient and crew friendly.

Self-cleaning smokeboxes, rocking grates, hopper ashpans and mechanical lubricators speeded up locomotive preparation and disposal. While still on the drawing board Ivatt had a full size mock-up built of the tank engine for inspection and comments by both footplate crew and maintenance staff.

Their recommendations were incorporated into the designs before construction started and resulted in enclosed tender cabs and inset bunkers providing comfort and good vision when running tender or bunker first. Additional features included tender and bunker access ladders, ample footplating in front of the smokebox and a specially designed box to house the fire-irons.

In spite of all this careful planning performance of the initial engines was disappointing, attributed mainly to draughting problems and the variable quality of available coal. Tests and experiments at both Derby and Swindon resulted in changes to the blastpipe and a reduction in the internal diameter of the chimney, modifications which resolved the engines' steaming problems.

The engines were now demonstrably economical, a pleasure to drive, capable of rapid acceleration and able to maintain high speed running in excess of 70mph. As proof of their success they provided the basis for the British Railways standard class 2 2-6-0s and 2-6-2 tanks with virtually no modification.

Construction of both classes was continued by British Railways and by 1953 there were 128 2-6-0s and 130 2-6-2 tanks, popularly known as "Mickey Mouses".

Seven 2-6-0s and four tanks have survived into preservation although at present most are either under overhaul or awaiting preservation.

54

London & North Eastern Railway

A1 & A2 PACIFICS [1947]

Arthur Peppercorn was responsible for the last of a famous line of LNER Pacifics, his A2 of 1947 becoming one of the most powerful express passenger locomotives to work on Britain's main lines.

BELOW LNER Peppercorn Class A1 4-6-2 locomotive 60131 "Osprey" at Leeds Neville Hill loco shed

THE A2'S

The initial design of the A2 Pacific was started by Arthur Peppercorn's predecessor, Edward Thompson not long before his retirement, but was completed and improved by his successor, Arthur Peppercorn, known universally as "Pepp".

Peppercorn redesigned the front end, added ideas of his own, and rectified a number of fundamental faults in the original Thompson design. The result was a more balanced and much better looking engine.

The class of 15 A2s, British Railways Nos. 60525–39, was completed at Doncaster by August 1948.

THE A1'S

In most respects, the A1 resembled its immediate predecessor. The principal difference was the driving wheel diameter, which, at 6 ft 8 in, was 6 inches greater and consequently necessitated a longer wheelbase. The arrangement of cylinders and valve gear was identical and the boilers were interchangeable.

The A1s became famed for their reliability. By 1961, the class of 49 had run 48 million miles, a figure unmatched by any other steam locomotive. All 49 engines were withdrawn and scrapped between 1961 and 1966.

A NEW A1 – 60163, TORNADO

To make up for the fact that no A1 Pacific had been saved for preservation the A1 Steam Locomotive Trust was formed in 1990 to build a brand new A1 Pacific. Costing £3m and taking almost 20 years to build, the new locomotive was successfully tested in November 2008 and formally unveiled at the National Railway Museum on 13th December 2008.

ABOVE Brand new steam locomotive 60163 Tornado on her first main line passenger journey to London, The Talisman, 7 February 2009

British Railways
STANDARD CLASS 7 BRITANNIA PACIFIC 4-6-2 [1947]

The Standard Class 7 Pacifics, referred to as Britannia Class after the name of the first engine, were Robert Riddles' first Standard Class design for British Railways. The guiding principles of the Standard designs were simplicity, use of established technology, ease of access for maintenance and incorporation of as many labour saving devices as possible. Simplicity and ease of access dictated outside cylinders without the complication of inside cylinders and conjugated valve gear. Accordingly the Britannia Class were Britain's first Pacific locomotives to be built with just two cylinders.

BELOW 70013 Storms the climb out of Sheringham, near to Upper Sheringham, Norfolk

Echoing his own work experiences the Standard designs drew largely on LMS design and construction practice but also incorporated the best features of other companies' locomotives that had been established in the 1948 Locomotive Exchange Trials.

In the case of the Britannias this meant boiler and trailing wheel designs borrowed from Bullied's Merchant Navy Class, weight saving construction techniques adapted from Bullied's Light Pacifics and, unusually in a modern express locomotive, a single chimney and blastpipe designed by S O Ell at Swindon.

The 55 engines of the class, built in three separate batches between 1951 and 1954, were spread around British Railways Regions and were generally well liked, especially on the Eastern Region where crews appreciated the low weight to power ratio on the restricted lines of the old Great Eastern Railway. The engines were less well liked by crews on the Western Region but this was largely a partisan reaction to the Britannias left-hand driving position compared to the traditional Great Western right-hand drive.

Competent and workmanlike, if a little unexciting, the Britannias never roused enthusiasts' passion in the same way as the pre-war A4 and Duchess Pacifics. Rather like the comparison between a run of the mill Ford Cortina and the performance, say, of an E-Type Jaguar.

Following a suggestion by the famous railway photographer Bishop Eric Treacy the engines were all named after historical British figures with the curious exception of No 70047 which never received a name.

ABOVE 70013, Oliver Cromwell, at Bressingham Steam Museum, UK

The engines were withdrawn between 1966 and 1968 and two have been preserved. No 70000 Britannia has been preserved privately and 70013 Oliver Cromwell, which achieved fame hauling the final Fifteen Guinea steam specials on British Railways, was preserved as part of the National Collection on its withdrawal in 1968. Both engines have now been restored and modified to permit to permit modern day main line running.

56

British Railways STANDARD CLASS 5 4-6-0 [1951]

BELOW BR Standard Steam Locomotive 73096 at Kidderminster SVR Station during the 2003 Autumn Steam Gala

Though bearing a strong superficial appearance to the LMS's Black Five the Standard Class 5 embodied a number of improvements, particularly in relation to crew comfort.

On nationalisation British Railways inherited three broadly similar mixed traffic 4-6-0s: the LMS "Black Fives", GWR Halls and Thompson's B1s from the LNER.

The new Standard Class 5 drew on all the best features of the inherited engines but, like all the Standards, looked like an LMS engine and bore, in particular, a striking resemblance to Stanier's Black Five.

But there were some significant changes and improvements, many aimed at operating efficiency and crew comfort. The new engines had self cleaning fireboxes, rocking grates and prior to production mock ups of the standard's cabs were produced specially to gauge the comments of footplatemen on layout, controls and instruments. The cabs of the

ABOVE Caprotti 5 73129 works a demonstration freight train across Butterley causeway

new engines were larger than the Black Fives, had padded seats for both driver and fireman with controls that remained in reach while they were seated and GWR style gauges which were easier to read than the LMS variety.

Other changes from the Black Five were larger driving wheels, 6' 2" instead of 6', Southern style clack valves instead of the leaky LMS top-feed, and a high running plate with a deep valance.

The Standard Class 5s were well liked by their crews and had a reputation for good acceleration, reliable adhesion and the capability of sustained high speed running.

172 Standard 5s were built between 1951 and 1957, 130 at Derby and 42 at Doncaster, and were allocated to all six regions of British Railways. 20 engines on the Southern Region inherited names from King Arthurs as these engines were withdrawn.

Five engines of the class, which had a woefully short working life of only 17 years on British Railways, have been preserved. Interestingly No 73096, based on the Mid-Hants Railway has already put in more years service there than it did on British Railways.

57

British Railways

STANDARD CLASS 4 4-6-0 [1951]

RIGHT 75027 at Horsted Keynes, on the Bluebell Railway

Constructed more lightly than their Class 5 shed mates the Class 4 4-6-0s were admirably suited to the secondary line duties for which they were designed.

Though similar in appearance to the Standard Class 5, design of the Standard Class 4 was actually inspired by the final H G Ivatt version of the LMS 2-6-4 Tank.

With the tank's leading pony truck replaced with a Standard Class 5 bogie the Class 4 was constructed more lightly than the Class 5 to give it universal access over all Britain's main and secondary lines.

A maximum axle loading of 17 tons 6cwt compared to the Class 5 figure of 19 tons 14cwt meant the Class 4s were widely used on cross-country lines and proved particularly useful on the former Cambrian Railway routes.

The 80 engines of the class (all built at Swindon between 1951 and 1957) shared the labour saving devices, self-cleaning fireboxes, rocker grates and hopper ash pans and the crew comforts associated with the Standard Class 5s.

Unlike the Class 5s which were allocated to all six British Railways' regions the class 4s were initially only allocated to three regions, Southern, Western and London Midland.

Competent if unexciting the Class 4s were perfectly matched to their secondary line duties with acceleration and speed appropriate to their stop-start duties.

Withdrawals started in 1964 but five of the class soldiered on to the last day of steam in August 1968. Six of the class have survived into preservation, five in fairly active use and one, No 75079, still waiting to be restored to working order.

58

British Railways

STANDARD CLASS 4 4-6-2T [1951]

Although superficially very similar to the 1945 2-6-4 tanks designed for the LMS by Charles Fairbairn, modifications required to ensure the new Standard engine would fit the British Railways universal L1 loading gauge resulted in a more handsome looking engine.

BELOW 80072 at Carrog Station

The 2-6-4 tanks, though designated mixed traffic (4MT), were primarily used on fast outer suburban passenger duties. At nationalisation the London Midland region was well equipped with ex LMS 2-6-4 suburban tanks whose evolution could be clearly traced back to Henry Fowler's design of 1927 with subsequent modernised variants designed by both William Stanier and Charles Fairbairn.

The Western Region was similarly well equipped with ex GWR Large Prairie 2-6-2 tank engines but on the Southern Region secondary steam hauled services were still largely in the hands of pre-grouping locomotives and there was pressing need for something more modern. To meet this requirement the original intention was to carry on building Fairbairn's 2-6-4 tank design as one of British Railways new standard classes but closer investigation showed that the engine would need considerable modification to meet the new country wide universal L1 loading gauge.

ABOVE 80064, Bluebell Railway

The main changes involved curving the tank and bunker sides inwards, giving the engines their attractive appearance compared to the more slab sided Fairbairn design, and providing smaller diameter cylinders, 18" compared to 19.625". To maintain the tractive effort of the Fairbairn engines the piston stroke was increased to 28" and boiler pressure raised to 225 psi.

155 of the engines were built between July 1951 and November 1956 with the majority, 130 of them, constructed at Brighton Locomotive Works. Initially delivered to all British Railways Regions except the Western the engines were commonly seen on the ex-London, Brighton and South Coast Railway routes of the Southern Region and the ex-London, Tilbury and Southend Railway lines of the Eastern Region.

Electrification and the introduction of diesel and diesel-electric multiple units saw rapid withdrawal of the engines from 1964 onwards, long before the end of their useful lives, with the last nine working members of the class being withdrawn at the end of steam on the Southern Region in July 1967.

An amazing 15 engines of the class have survived into preservation, largely explained by the 14 members that ended up in Woodham Brothers Barry scrap yard from where they were slowly rescued and restored.

59

British Railways

STANDARD CLASS 8 PACIFIC No 71000 DUKE OF GLOUCESTER [1954]

Although Robert Riddles was keen to add an 8P Pacific to his roster of Standard Locomotives the Railway Executive rejected his proposals as there were sufficient 8P Duchess Pacifics to meet demand on the West Coast Main Line. However the destruction of Pacific 46202 Princess Anne in the 1952 Harrow and Wealdstone rail disaster provided the opportunity to replace the loss with a single 8P Pacific prototype.

BELOW 71000 Duke of Gloucester at Victoria Station

Riddles original proposal was to build a larger version of the Britannia Pacifics, maintaining the simplicity of just two cylinders. Unfortunately to achieve the 8P power output the cylinders would have been too large for the British loading gauge forcing the reluctant compromise of a third inside cylinder. To avoid the complication of conjugated valve gear for the inside cylinder Riddles elected for British Caprotti valve gear, a British modification of the Italian design featuring poppet valves driven by a rotary cam.

The characteristic of Caprotti valve gear is precise control of steam admission with improved exhaust flow and boiler drafting. For maximum efficiency the British Caprotti company recommended that the valve gear, which generates a fierce exhaust, should be used in conjunction with a Kylchap blastpipe.

Unfortunately, for cost saving reasons, the engine was instead fitted with a standard Swindon double chimney with too small a choke area between blastpipe and chimney to cope with the pressure of the Caprotti exhaust. This flawed choice produced an engine that was a poor steamer with a reputation for greedy consumption of both coal and water. Midland Region crews used to working on Duchess Pacifics hated No 71000 Duke of Gloucester and much of its short 8 year working life was spent on less demanding duties such as the Mid-day Scot, weifgt restricted to eight coaches.

But the story has a happy ending. Rescued from Woodham Brothers scrapyard in 1974 the locomotive was restored back to working order over the next 12 years. Restoration included fitting the correct Kylchap blastpipe and correcting errors to the original design of the firebox and ashpan. Subsequent Lottery funded improvements between 1997 and 2004 have produced a formidable locomotive frequently seen on mainline tours and probably now performing just the way Riddles imagined she should.

ABOVE 71000 DUKE OF GLOUCESTER of Bury MPD at Bury South on the Down Main line on the East Lancashire Railway

60

British Railways STANDARD CLASS 9F 2-10-0 [1954]

BELOW British rail standard 2-10-0 reposes in the sidings at Horestead keynes

Nothing underlines better some of the wrong-headed aspects of British Railways 1955 Modernisation Plan than the life histories of the 251 Standard Class 9F 2-10-0 freight locomotives. Constructed between 1954 and 1960, the 9F was the last and best regarded of Robert Riddles' 12 standard class locomotives designed for Britain's nationalised railways. But with withdrawals taking place as early as 1964 some of the locomotives' working lives were truncated to just five years, compared to the 20 or 30 years of work normally expected from engines such as this.

The need to haul heavy freight trains up to 900 tons at speed on round trips like the merry-go-round service between colliery and coal fired power station was the rationale behind the construction of the 9F 2-10-0s. Although Riddle's initial thoughts for the design were for a 2-8-2 engine he finally drew inspiration from the rugged simplicity of the 2-10-0

"Austerities" he'd designed for use by the War Department during World War II.

Like all the standard designs the 9Fs relied on simplicity and proven technology while incorporating all the modern operational benefits like roller bearings, self-cleaning smokeboxes, rocking grates, mechanical lubricators and cab design that placed the emphasis on crew comfort.

After some very minor teething problems the engines were quickly established as successful freight locomotives and did much to improve British Railways freight operations during the last decade or so of steam. And though discovered almost by accident they were also excellent passenger engines, deputising on summer specials and proving particularly effective on the steeply graded Somerset and Dorset line in the final years before its closure.

As freight engines all the class were painted unlined black with the sole exception of No 92220 Evening Star illustrated here. Evening Star was the last locomotive built at Swindon and to celebrate the event the engine was finished in fully lined out Brunswick green with a GWR style copper capped chimney. The name, Evening Star, so appropriate and redolent of endings, was chosen from a completion held among Western Region staff and unveiled in a ceremony at Swindon on 18th March 1960.

Evening Star is preserved as part of the National Collection but another eight engines have survived: seven rescued from Barry scrap yard and one, No 92203, purchased privately on withdrawal by David Shepherd, the famous railway and wild life artist.

ABOVE Standard Class 9F No 92219 awaiting restoration at Swanwick on the Midland Railway Butterley

The pictures in this book were provided courtesy of the following:

WIKIMEDIA COMMONS

Design & Artwork: ALEX YOUNG

Published by: G2 ENTERTAINMENT LIMITED

Publisher: EDWARD ADAMS & JULES GAMMOND

Written by: IAN MACKENZIE

Printed in Dunstable, United Kingdom

91222428R00074

Contents

INTRODUCTION

To mark its status as the last steam locomotive to be built by British Railways, standard 9F 2-10-0 freight class engine no 92220 was named Evening Star at a special ceremony at Swindon Works on the 18th of March 1960.

Correctly signifying an ending, the name, chosen from a competition run in the BR Western Region Staff Magazine, provided an apt foil to Morning Star, a broad gauge locomotive built in 1839 and one of the Great Western's earliest engines.

Marking its historic importance, the engine was turned out in fully lined Brunswick green livery with a copper-capped chimney and brass fittings in contrast to the utilitarian unlined black livery of all its class mates.

And as the final milestone in a century and a half of British locomotive engineering it was right to acknowledge the importance of this historic event.

Along with Evening Star the other 59 engines in this Little Book of Classic Locomotives also represent important milestones along one hundred and fifty years of British locomotive development. Demonstrating significant technological advance the inclusion of some locomotives is obvious but space has been made for the humble freight engines and suburban tanks that played such an important role in Britain's steam heritage.

To avoid screeds of technical jargon and tables of superfluous data the narrative on each of the 60 locomotives has been limited to less than 400 words, just the salient features about the engine, its designer and its place in history.

There are omissions, argument perhaps for a second volume of Classic Locomotives, but with just three exceptions, all the engines included are still extant and can be seen as static exhibits in museums, working on heritage railways or even running on mainline specials.

Presented chronologically, the book

ABOVE 34046 Braunton at Barry

starts with Robert Stephenson's Planet and Patentee class of locomotives whose design established future locomotive development more clearly than either Locomotion or Rocket and which also underpinned Robert Stephenson and Co's burgeoning export business. Germany's famous first engine, Adler, was delivered by Robert Stephenson and Co as a kit of parts to the Bavarian Ludwig Railway in 1835.

Germany has a working replica of Adler, constructed in 1935, and a working replica of Planet, built in 1992, can be seen at Manchester's Museum of Science and Industry.

Nearer the end of our chronological journey the word replica inadequately describes A1 Pacific No 60163 Tornado which was first steamed in 2008 after almost 15 years of construction. Faithfully preserving the appearance of Peppercorn's A1 pacifics, Tornado incorporates all the engineering improvements developed over the past fifty years. Not so much a replica, more a modern twenty first century steam engine.

Costing over £3 million Tornado underlines Britain's enthusiasm, obsession even, for steam railways: an obsession which supports almost 200 active heritage sites and associations. Other countries have steam museums and heritage sites, but nothing on a scale that compares to Britain. Why is that? Is it pride in an engineering tradition that started in Britain or simple nostalgia for a bygone age, hankering back to good old days which are perceived as more generous, less demanding and fairer than the modern age?

One explanation is the sheer speed of change on Britain's railways in the decade up to 1970. Driven by both the 1955 Modernisation Plan and Dr Breeching's Reshaping of British Railways, that decade saw the scrapping of 15,000 steam engines, the rail network reduced by a third and the closure of over 2,000 stations and virtually all rural goods yards.

Even after nationalisation in 1948 and in spite of some livery changes, the railways familiar to train-spotters in the 1950s had changed little since the 1920s. So the rapid and almost traumatic change to this cosily familiar scene quickly galvanised steam enthusiasts to save what they could.

And the resources were there: disused branch lines, withdrawn steam engines and redundant coaching stock. Quick resolution of legal wrangles with British Railways soon saw volunteers running steam trains on faithfully re-enacted branch lines, their station platforms carefully staged with all the features beloved of railway modellers: the antique luggage, porters' barrows, milk churns and enamel advertisement signs.

So prolific was the growth of these heritage sites, including famous early pioneers such as the Bluebell, the Severn Valley, the Keighley and Worth Valley, the Watercress Line and the Kent and East Sussex, that at one stage it seemed there would be a famine of locomotives to preserve. But then, in a curiously British and serendipitous quirk of history, there was the Woodham Brother's Barry scrapyard.

Almost a quarter of preserved steam engines running in Britain today were rescued from Barry, some after mouldering in the South Wales sea air for two decades and then taking a further twenty or even thirty years to be fully restored to working order. Looking at pictures of engines at Barry prior to rescue, their fittings removed and steel plate reduced to a fine rusted lace, you have to wonder how much of the original engine exists.

Because of its geographic location many of the withdrawn engines that ended up at Barry, including British Railways' standard classes, came from the Sothern and Western Region. Ex Barry engines include Castles, Kings, Standard Class 4s and 5s and, particularly, Bullied's light Pacifics. From a class that originally totalled 110 engines no fewer than 20 have been rescued from Barry. Were it not for Dai Woodham no Bullied light Pacifics would have been preserved.

By contrast engines from the Eastern and North Eastern Regions fared less well, most being cut up soon after withdrawal. For example, the LNER's mixed traffic B1

4-6-0, a class which totalled 410 engines, is today represented by just two preserved locomotives, there is just one preserved NER Q6 0-8-0 from a class of 120 engines and those enthusiasts who wanted to see a Peppercorn A1 Pacific saved for posterity had to build their own.

A number of British Railways Standard class engines have also been rescued from Barry. After nationalisation in 1948 British Railways built almost 3,000 new steam engines including 999 Standard Class engines designed by Robin Riddles, the first Chief Mechanical engineer of the new railway. At a time when continental railways were already scrapping steam and investing heavily in electric traction it seems curious that British Railways chose instead to invest in steam traction. Although today that choice may seem wrongheaded the strategy was actually very logical. After the Second World War Britain's financial state was even more parlous than it is today. But it had the necessary skills and resources to build steam engines that ran on indigenous coal and couldn't risk straight away investing wholesale in new technology that would have involved importing know-how as well as fuel for diesel powered engines.

With hindsight what does seem questionable was the almost unseemly speed of steam locomotive scrapping. Some Standard class engines built in the late 1950s and 1960, like Evening Star, had a working life of little more than five years.

And, finally, what are the three exceptions in this book that haven't survived to the present day. They are John McIntosh's 1896 Dunalastair 4-4-0 class for the Caledonian Railway, Charles Bowen Cook's flawed 1913 Claughton class 4-6-0 design for the London and North Western Railway and the Great Western Railway's County class of 1945. But read the narrative for each of these engines and you'll see that by a quirk of history and the enthusiastic intent of modern day preservationists have ensured a sort of life after death for even this trio.

So perhaps you should regard this volume of Classic Locomotives as a latter day Ian Allan Spotting Book and see just how many of the 60 engines you can cop.

Liverpool & Manchester Railway

PLANET 2-2-0 [1830]

BELOW
Side and end views of a locomotive engine similar to the Planet (Liverpool and Manchester Railway) engraving by William Miller 1832

Planet, delivered in 1830 was the ninth locomotive built for the Liverpool & Manchester Railway by Robert Stephenson & Co. Although separated by just a year Planet represented a considerable improvement over Rocket, the Stephenson's first engine for the railway.

The main changes which distinguished Planet from Rocket and which were largely intended to improve the ride and stability of the engine, was a switch in the position of the driving wheel, from Rocket's 0-2-2 arrangement to 2-2-0, and a lowering of the centre of gravity by moving the cylinders inside the frames below the smokebox. Other improvements established features familiar on all subsequent steam locomotives: heavy sandwich type frames, substantial buffer beams and buffers and a water surrounded firebox integrated with the boiler.

These improvements produced a machine still primitive looking to modern eyes but, compared to its rickety looking predecessors, more recognisable as a steam locomotive. Planet also established itself as an early "standard", forming the basis of the Stephenson's burgeoning export business and with the name being used as the "type" description for all subsequent 2-2-0s employing Stephenson's valve gear.

PATENTEE 2-2-2

In daily use the shortcomings of the short wheelbase four-wheeler Planet type became apparent: a front-to-back pitching motion, instability at speed and the frustrating lack of effective reversing gear. Shortcomings quickly addressed with Robert Stephenson's 1833 locomotive Patentee. With a name implying the importance of patenting new locomotive developments Patentee also established another important class type, the inside cylindered 2-2-2, familiar on Britain's railways for the next three decades with some examples remaining in use to the end of the nineteenth century.

Patentee was immediately distinguishable from Planet with its third set of rear wheels, able to support a much larger and heavier firebox. Other less obvious changes were improvements in boiler construction and valve gear and the introduction of a counter-pressure steam brake, an idea somewhat ahead of its time.

With its extra set of wheels and longer wheelbase the Patentee type ran more steadily at speed, were less likely to derail and saw an increase in engine power with their bigger boilers and fireboxes. The additional axle also spread the engines' heavier weight more evenly but also minimised the individual axle loadings, an important consideration in view of the weight restricting fish-bellied cast iron rails in use during the 1830s.

ABOVE A working replica of Robert Stephenson's's 1830 locomotive Planet which ran on the Liverpool and Manchester Railway. Seen here at the Manchester Science Museum

2

Liverpool & Manchester Railway

LION [1838]

BELOW Lion built by Todd, Kitson and Laird, who used a conventional 0-4-2 wheel arrangement with inside cylinders.

As one of four relatively mundane "luggage engines" built in 1838, Lion's fame and survival to the present day relies on some quirky turns of fate and not a little serendipity.

Though built less than ten years after Rocket, Lion presents an altogether more robust appearance and already displays familiar features of nineteenth century locomotives: running plate, wheel splashers and buffer beams together with inside cylinders and altogether more substantial double frames.

Surprisingly, during this era of rapid locomotive development and early retirement, Lion remained working for the Liverpool and Manchester Railway, by then the London & North Western Railway until 1859.

Lion was then sold to the Mersey Docks and Harbour Board for £400 and then spent several years as a dockside shunter. In the late 1860s Lion was taken off the rails and mounted as a static pumping engine.

Still at work over 60 years later, Lion was discovered by a visitor to the docks, C.W.Reed, in 1927. In 1928 Lion was finally replaced by an electric pump and in 1929 the engine was presented to the Liverpool Engineering Society. The Society persuaded Henry Fowler, the CME of the London Midland & Scottish Railway, to overhaul the engine so that it could take part in the centenary celebrations of the Liverpool and Manchester Railway, which it duly did.

Then began Lion's career as a film star: in Victoria the Great (1937), The Lady with the Lamp (1951), a film about Florence Nightingale and most famously as The Titfield Thunderbolt in 1953. Immortality was assured.

Now a static exhibit at the Museum of Liverpool Life, Lion was last overhauled in 1979 ready to appear in the cavalcade of locomotives at the Rainhill 150 event in 1980 and continued to be steamed at a variety of events up to 1992.

ABOVE Locomotive Parade, Rainhill 1980: 'Lion' was one of four locomotives built in 1838-9 for the Liverpool and Manchester Railway, and is the sole survivor of the four

3

LBSCR

TERRIER TANKS 0-6-0 [1872]

RIGHT TOP LBSC AiX No. W8 Freshwater at Kingscote station,

BELOW Bodiam at Rolvenden on the K&ESR

Although only fifty Stroudley A1 Terriers were built between 1872 and 1880 ten of these remarkable little engines have survived to the present day.

Initially built for passenger services on the LBSCR's south London line linking Victoria and London Bridge, the engine's design needed to accommodate a light weight combined with rapid acceleration.

Contemporary accounts suggest that much of the south London track, constructed of lightweight wrought iron rails, was in such poor condition that it could not be trusted to support a locomotive weighing more than 25 tons.

Stroudley appears to have relished the challenge presented by these design requirements and produced an engine weighing only 24 tons 7cwt whose power and acceleration belied its compact design.

The first six engines were delivered in the

autumn of 1872 and promptly astonished everyone with their performance acquiring, straightaway, the very appropriate Terrier nickname.

In 1911 Lawson Billinton, the LBSCR's newly installed locomotive superintendent, decided to extend the lives of the company's Terriers with new boilers and larger cylinders. The altered engines were designated class A1x.

The remarkable power to weight ratio of the Terriers explains the survival of so many engines of this class and all the remaining Terriers finally retired by British Railways in 1963 have ended up in preservation.

ABOVE Stroudley Terrier 0-6-0 "Boxhill" at the National Railway Museum, York, UK

4

The Great Eastern Railway

Y14 (LNER J15) 0-6-0 [1883]

T.W.Worsdell's Y14 six coupled freight engine first introduced in 1883 proved so reliable that it continued to be built over the next 30 years.

BELOW
J15 65462

The opening of the Great Eastern / Great Northern joint line linking March, Spalding, Lincoln and Doncaster meant that the GER's normal goods traffic of agricultural products and fish was suddenly expanded by Yorkshire coal bound for London.

This change demanded fresh freight locomotives and Locomotive Superindent T.W.Worsdell, who had joined the Great Eastern in 1882, made the design of a new goods engine his priority.

The first engine of the new 0-6-0 class, Y14 number 610, rolled out of Stratford Works in 1883. Modestly proportioned, the new engines quickly earned the nickname "Little Goods" although this completely belied the capable and powerful nature of these machines. By 1892, 229 Y14s had been built and further